DRIVING TO MARS

ALSO BY WILLIAM L. FOX

Nonfiction

Terra Antarctica: Looking into the Emptiest Continent
In the Desert of Desire: Las Vegas and the Culture of Spectacle
Playa Works: The Myth of the Empty
The Black Rock Desert
Viewfinder: Mark Klett and the Reinvention of Landscape Photography
The Void, the Grid, & the Sign: Traversing the Great Basin
Driving by Memory
Mapping the Empty: Eight Artists and Nevada

Poetry

Reading Sand: Selected Desert Poems, 1976–2000
One Wave Standing
silence and license
geograph
Leaving Elko
Time by Distance
21 and Over
The Yellow Pages
First Principles
Monody
Election
Trial Separation
Iron Wind

DRIVING TO MARS

William L. Fox

SHOEMAKER & HOARD

Definition on page 69, "Areo-" The Oxford English Dictionary. 2nd ed. 1989. OED Online. Oxford University Press. 15 Apr. 2006—<http://dictionary.oed.com/cgi/entry/50011760>. All gallery illustrations used with permission and credited as noted. Photos by Peter Essick courtesy of Peter Essick/Aurora. Every attempt has been made to secure permissions. We regret any inadvertent omission.

Library of Congress Cataloging-in-Publication Data

Fox, William L., 1949–
Driving to mars / William L. Fox.
p. cm.
Includes bibliographical references and index.
ISBN-13: 978-1-59376-111-0
ISBN-10: 1-59376-111-2
1. Mars (Planet)—Geology. 2. Mars (Planet)—Exploration. 3. Haughton Crater (Nunavut)—Discovery and exploration. I. Title.

QB643.G46F69 2006
559.9'23—dc22

2006008562

Cover design by Gerilyn Attebery
Interior design by Megan Cooney
Printed in the United States of America

Shoemaker & Hoard

An Imprint of Avalon Publishing Group, Inc.
1400 65th Street, Suite 250
Emeryville, CA 94608
Distributed by Publishers Group West

10 9 8 7 6 5 4 3 2 1

To the Inuit

&

the astronauts

CONTENTS

PREFACE
At the Edge of Endurance

MAY 2004. The only motion on the somber Martian plain is the halting progress of what looks like a squashed golf cart carrying a camera mast and a jointed mechanical arm. The *Opportunity* rover creeps forward a few inches, its six wheels rocking gently up and down over small pebbles, then stops. Several of its nine cameras traverse the view ahead, scanning the horizon and mapping obstacles in the next few feet. After twenty seconds the rover resumes its pedantic pace, heading slightly to the left to avoid a rock the size of a football. Ten more seconds of movement at just under one inch per second. Pause for twenty seconds, resume for ten. Pause, resume, then full stop.

Directly in front of *Opportunity* is the edge of its goal, the Endurance Crater, which the rover has taken six weeks to reach from where it landed a half-mile away. Named after Sir Ernest Shackleton's legendary ship, which was crushed by ice in the Antarctic during the winter of 1915, the crater is 490 feet across and 66 feet deep. Eleven light minutes away on Earth the engineers teledriving the rover at the Jet Propulsion Laboratory in Southern California are staring at a stratigraphic history of Mars. Layers of bedrock exposed on a cliff across the crater look like they could be either basalt from volcanic action or sandstone laid down by an ancient sea.

For the next several weeks the rover will gingerly make its way by remote control around the crater's quarter-mile perimeter, the engineers trying to avoid tipping over the rim by accident. The gentlest slope of the crater wall is only around twenty degrees, but the rover had almost gotten stuck a few weeks ago in the sand of an even milder incline as it tried to climb out of a smaller crater. The scientists aren't ready yet to make the decision whether or not to send the vehicle into Endurance. Once inside, the rover could be doomed to spend its remaining days on the dusty floor, as decisively trapped as Shackleton's ship was when it was nipped in the ice. On the other hand, it would be an unparalleled chance to examine that cliff, to determine whether or not an ocean existed here. An ocean that might have contained life.

The equatorial plains of Meridiani Planum sit within an area the size of Oklahoma that is sprinkled with small "blueberries" of hematite, gray pebbles of iron oxide that account for the ground's uncharacteristically dark color. It's good driving terrain because the pebbles and rocks are thinly scattered atop a fine-grained sand more than nine hundred feet deep. The temperature of the almost imperceptible atmosphere has been cooling from its summer high of 32°F and every night now dips below minus 50, dropping ever closer to the estimated winter minimum of minus 148°F. The vehicle's battery and electronics sit inside an insulated box, which is heated by a slug of rapidly decaying plutonium dioxide the size of a pencil eraser. In theory the rover can withstand external temperatures down to minus 157°F. At this point, however, it's anyone's guess whether or not it can maintain power during the five-month-long Martian winter.

While being flown from Earth *Opportunity* traveled at speeds in excess of 17,400 miles per hour and crossed 283 million miles between its launch on July 7, 2003, and its landing on January 24, 2004. On the relatively smooth surface of Meridiani Planum its land speed record is 462½ feet in one day. Both *Opportunity* and its twin rover, *Spirit,* which is on the other side of the planet, have been averaging only a little more than forty yards per day in their travels. While the former has been creeping toward the crater, *Spirit* has now broken the long-distance record for travel on Mars, having passed the one-mile mark on its way to the Columbia Hills. Both rovers have accomplished their primary three-month missions and are now operating on extended status, which will last until they can no longer generate enough power to function, or one of the separate electrical motors driving each of their six wheels wears out—or until they are sent on a suicide mission.

The rovers aren't looking for life, but for evidence that the conditions necessary for life as we understand it once existed on the planet. And that means finding signs of water. *Opportunity*'s specific and primary objective was to examine the local hematite to see if it was caused by the presence of water, and the rover succeeded beyond expectations. The shallow sixty-six-foot-diameter crater into which the lander had bounced contained a few inches of exposed bedrock. Embedded in it were many of the spherical BB-size blueberries, which spectral analysis showed to be the kind of hematite that forms in wet conditions. In addition, the rocks contained enough signature minerals to confirm that saltwater was once present. The cameras also spotted what turned out to be cross-bedded layers of sediment, formed under flowing water. The stratigraphy meant that the crater had once sat in what may have been a salt flat or playa, the type of intermittently wet and dry lakes found in places such as Nevada. This was the first time that ground truth had been established for water on the planet.

Now the rover has reached a much larger crater, and the exposed bedrock on one of its outcrops is more than thirty feet high. The history it represents could be billions of years old. Already, however, the rovers are covered in a thin coat of dust. The orangish particulates, combined with the inexorably decreasing sunlight of the approaching Martian winter, have cut the efficiency of the solar panels by 30 percent. The engineers have to balance how much power *Opportunity* can collect and store with the increasing amounts of energy needed each night to maintain heat inside its insulated box. Added to that is the decision by the scientists whether or not the science to be gained is worth the risk of losing the rover inside the crater.

If a human were here, he or she could walk down inside the crater and accomplish in one eight-hour shift what the rover—essentially a robotic geologist—will take more than two months to do. While *Opportunity* painstakingly probes the rim of the crater, and the engineers and scientists on Earth debate the fate of the rover, another group prepares to practice Mars on Earth.

The sun is a feeble disk sinking low on the horizon of Mars, but back on Earth in the Canadian High Arctic it's spring, and under lengthening days of sunlight the sea ice is beginning to break up along the legendary Northwest Passage. Dozens of scientists and engineers from NASA, the Canadian Space Agency, and various universities are preparing to converge in July on Resolute, the northernmost settlement on the passage. From there they will fly over a channel to the east and land on Devon Island, its dusty and almost sterile landscape one of the closest approximations on Earth to the Martian surface. At the edge of a thirty-eight-million-year-old impact crater twelve miles across they will break out tents and sleeping bags, and go through the annual rituals of reestablishing camp for the NASA Haughton-Mars Project (HMP). The portable generators will rumble to life, a large satellite antenna will be pointed at the horizon, and a data pipeline extended to the outside world. Once people are settled in, they will roll out the four-wheel all-terrain vehicles, or ATVs, from one of the three large tents left up during the winter, and they'll pull the tarp off an ungainly orange Humvee. These vehicles will serve as the roving test beds to be driven across the world's largest uninhabited island.

The HMP camp is devoted to conducting field science and developing protocols for how humans and robots will explore Mars. Three-quarters of a mile away stands the two-story white cylinder of the Mars Society's Flashline Arctic Station, the outpost of a private organization that has a different goal in mind, the colonization of the Red Planet. Devon Island isn't just a fine analog for Mars, but also for the competing political and economic interests that, along with science, have driven the engines of exploration.

Each morning after breakfast in the large kitchen tent, people will check the progress of the rovers on the Internet, scanning images from the Red Planet. Then they'll pull on their parkas and gloves to slip outside in temperatures hovering around freezing. Teams will load up the ATVs and Humvee with instruments and provisions, and head out for features that resemble those on Mars—the crater, water gullies, meltwater channels, and caves with perennial ice. One group will launch remote-controlled unmanned aircraft to scan the terrain, while another will take the Humvee to attempt a traverse north to the cliffs along the coast.

A third group will drive down into the crater with the upper half of a concept suit that will test what a scientist might wear on Mars. One of the camp's geologists will squeeze carefully inside its torso and helmet while wearing a heads-up computer display unit. Then she will walk slowly around the blasted landscape with a rock hammer and a long-handled grabbing stick. She will be given elementary tasks to perform—picking at a cliff face, looking at rocks, examining a fossil. The results will be relayed back to the NASA Ames Research Center at Moffett Field, a former airbase in California near Mountain View some 2,800 miles away, where engineers will time her actions in comparison with a robotic model. And where they daydream what it will be like when a robotic rover and a human scientist drive together into a crater on Mars.

CHAPTER ONE
Passages

THE FIGURE ahead of me kneels on his snowmobile as we race across the smooth sea ice at forty miles an hour. Paul Amagoalik, an Inuit from Resolute Bay, is casual yet alert, the hood of his knee-length black parka pulled up against the minus 5°F windchill. Half a mile to our left rises the steep south shoreline of Cornwallis Island; a hundred yards to our right sparkle the open blue waters of Barrow Strait, part of the Parry Channel and the main course of the Northwest Passage. The contours of the island are blurred by a gathering ground blizzard of spindrift that hasn't reached us yet. The ocean is still in sunlight, whitecaps sparkling. It looks like a day at the beach, but the briny Arctic water can run as cold as 29°F, three degrees below freezing.

By historical accounts, the strait shouldn't be open this early in the year, which means the currents are warm, maybe one or two degrees above freezing. It's only late April, and in previous years the passage sometimes never opened up all the way even by August. During the last decade, however, the ice has been moving out progressively earlier and polar waters have been open even at the pole itself, nine hundred miles to the north. Two summers ago a Canadian Mountie sailed a sixty-six-foot aluminum patrol boat through the passage in only twenty-one days. In the next ten or twenty years the Northwest Passage may become a viable route for international commerce, but it spells disaster for people traveling locally in the High Arctic—not to mention the polar bears, seals, and walrus who live on the ice. For most of the year the sea ice provides the shortest and safest route from one place to another, exactly what Paul and I are seeking so that Pascal Lee, the leader of NASA's Haughton-Mars Project, can drive a Humvee east from Resolute across the still-frozen twenty-three-mile-wide Wellington Channel to Devon Island.

As we attempt to navigate around the southeastern corner of Cornwallis to where the channel runs left and north off the strait, we're being forced slowly back toward land by the encroaching water. Paul scans the pressure ridges ahead of us, looking for a way to thread ourselves through the eight-foot-high barrier of upthrust ice. I'm enjoying the burst of speed and the intense blue of the polar waters, which is the most cheerful color I've seen in the landscape after several days of low gray clouds hanging over the snowy island.

Paul's back suddenly straightens. He sits down, the whine of his engine drops abruptly in pitch, and he swings a tight circle back the way we came. He looks over and gestures urgently. I back off the throttle and follow his track as carefully as I can. The ice has gone dark gray beneath us, and the surface is sickeningly elastic. We've been riding on ice six to eight feet thick, but here it has been hollowed out to only a few inches. Were I to stop, I could count individual bubbles rising underneath me from the warm current that's eroding the ice. It doesn't matter if the water is a degree or two colder or warmer than freezing. Either way, if we fall through, we die. I keep circling back very slowly.

Snowmobiles are an efficient way to travel over the frozen ocean. Unlike sleds pulled by dogs or people, they can cover huge distances in a short time, and Inuit all across the North American Arctic now use them for hunting and transportation. They are as ubiquitous among the two hundred people who live in Resolute as dirt bikes are in the small desert towns of California. Unfortunately, with a passenger and gear they weigh around 850 pounds each. A mass that size vibrating ahead at forty miles per hour creates a shockwave around the vehicle that will crack thin ice in concentric rings around you. If I go too fast or too slow, I'll quickly join the bubbles under my track. I match my pace to Paul's and hope we make the white ice just ahead.

Paul stops, pushes back his hood, and I pull up beside him on reassuring opaque ice. Our engines gutter in the cold and his shoulder-length black hair stirs in the breeze. "That was not good," he observes, his wispy mustache lifting with a quick grin.

I shake my head, shrug, smile back at him. "Looks like we'll have to stay closer to shore." He nods, puts the hood back up, and points us toward the pressure ridges we were hoping to avoid.

Devon and Cornwallis islands sit within the archipelago of the Canadian High Arctic, a labyrinth it took more than three hundred years to chart. The process of exploration, which started in the late sixteenth century, claimed hundreds of lives from scores of expeditions. The most obvious lure was the promise of a northwest passage across the top of North America, in theory a series of interconnected bays and straits that would lead westward from the Atlantic Ocean to the Pacific. All the way up through the mid-nineteenth century, Dutch and English trading companies had hoped that such a passage would provide them a shortcut to Asia, an alternative to the southerly routes dominated by the

Spanish and Portuguese. Beginning in 1817 the British Admiralty sent a series of expeditions into the ice-choked archipelago to discover a route through it. John Ross, Edward Parry, and John Franklin were among the explorers to probe westward into the maze. In 1845 Franklin led the most well-prepared polar expedition ever mounted into Lancaster Sound just east of Resolute and south of Devon Island—and disappeared.

The Northwest Passage does exist, and its elusive channels were mapped by 1859, in no small part due to information collected painstakingly by the forty subsequent expeditions searching for Franklin and his two ships, but a transit under sail was not completed until 1906. Roald Amundsen, a Norwegian who had grown up on skis in his native country, wisely adopted the clothing and survival skills of the Inuit—including dogsleds and a boat with a shallow draft—versus trying to overcome the Arctic with European technology, the strategy of most other polar explorers. It took him three years to link the passage together, but he succeeded by living within the constraints of the environment, instead of trying to overrule them. The lessons he learned in the Arctic served him well; in 1911 he used dogsleds at the opposite end of the planet to win the race to the South Pole. Robert Falcon Scott tried it in a more traditional English manner with his man-hauled sleds, but a month after Amundsen. He and his four companions died of cold and starvation on the way back.

Because it was understood by the mid-nineteenth century that the Northwest Passage was almost always choked by ice rotating down from the Arctic Ocean, it had lost its allure as a trade route even before Franklin disappeared, much less before Amundsen's traverse. Franklin's voyage had more to do with national pride, and the continued employment of naval officers necessary to maintain a viable military force in between wars, than it did with economics, although a shortcut to Asia avoiding the passage around Cape Horn had long been sought.

John Cabot, an Italian expatriate explorer living in England, was sent by Henry VII across the Atlantic to seek a passage as early as 1496, but by 1524 King Charles V of Spain was pondering another solution: He commissioned a study to see if a canal could be dug across the Isthmus of Panama. In 1855 the Panama Railroad began transshipping goods from the Atlantic to the Pacific across the fifty-mile-wide stretch of land, and when the Panama Canal was finally opened in 1914, it cut more than 7,800 miles off the route from New York City to San Francisco. Nowadays up to sixteen thousand ships make the day-long transit each year, approximately 5 percent of the world's trade.

The Northwest Passage, however, is rapidly becoming a more attractive proposition. Ten percent of the world's ships are already too large to fit through the aging canal, and the northern alternative offers a shortcut of more than five thousand miles to shipping between Europe and Asia. A government study shows that if Panama does not enlarge the canal by 2015 it will be obsolete; but the challenges are formidable. The government would have to replumb the watershed of virtually the entire country to provide enough water for the locks to operate on a larger scale; no one has any idea what to do with the billions of cubic yards of dirt that will have to moved; and there's the question of how to raise the $12 billion the expansion will cost. The United States, China, and Japan are the largest users of the canal, each of them well positioned geographically to take advantage of a northerly route, should one open.

In particular, oil companies would love to have a more direct way to ship Alaskan and undersea Arctic crude oil to refineries in northern Europe as the North Sea fields play out. In 1969 the double-hulled supertanker *Manhattan* made it through the Arctic passage on a successful test run, although not without paying a price. After breaking through fifteen-foot-thick sea ice and punching through forty-foot-high pressure ridges, it had numerous holes in its outer hull, and the oil companies drilling in Alaska's North Slope decided to build a pipeline instead, a pipeline that is now, like the canal, aging and a stationary target for terrorists. Since then the Arctic ice has thinned by more than 40 percent.

Open water notwithstanding, sailing through the Northwest Passage is still a tricky navigational proposition. Today we have maritime charts and topographical maps of the archipelago, aerial photos of the entire region, and constant satellite images that not only picture the clouds, but pierce through them with radar to see the extent of the sea ice underneath. We also have the Global Positioning System, or GPS, with which we can triangulate our position on ice or ground within a few yards. But the map is famously not the territory, and a satellite image is only a picture, not the place itself. All the remote sensing in the world cannot yet tell you what the exact sailing (or driving) conditions are from day to day, whether you're the captain of a supertanker or a scientist trying to get a large vehicle to Devon Island. And that's why Paul and I are out here hauling around on the sea ice.

Every July since 1997 scientists and engineers interested in exploring Mars have been flying into the dirt strip at Resolute on the specially equipped Boeing 727 passenger jets that serve the town, then transferring over to either helicopters or that workhorse of deep field aviation, the de Havilland Twin Otter aircraft. The powerful little planes, which reputedly can land and take off inside a football stadium, ferry fuel drums, ATVs, food for weeks, and up to a dozen passengers at a time to the site of their activities, the Haughton Impact Crater a hundred miles to the east.

Devon Island is the largest uninhabited island in the world, almost twenty-five thousand square miles of desolate rock and ice that sit between Baffin Bay and the Arctic Ocean. Its polar desert is home to a handful of scattered mammals and two dozen species of birds. Polar bears traverse its rocky slopes, where less than 3 percent of the ground is covered by vegetation, and the landscape is a dendritic maze of meltwater runoff channels. During the summer the island's central plateau—where the crater sits—is cold, dusty, and incessantly windy, a splendid terrestrial analog for the Red Planet. Once there, the visitors set up camp for the duration of the one-month-long summer that the Arctic offers at latitude 75° north.

The members of NASA's Haughton-Mars Project (HMP) drive around on ATVs as if they were riding rovers on the far planet. The project, under the leadership of Pascal Lee, a thirty-nine-year-old planetary scientist, is funded by a combination of government and private sources interested in the future of Martian exploration. It is an international and multidisciplinary effort that investigates the geology and biology of the crater—which is enough like Mars to provide clues about how to do science and look for life there—and that studies exploration hardware and techniques that might be used on the Red Planet by humans and robots. The engineers test out prototype pressure suits, communication links, and an experimental greenhouse built to grow tomatoes whenever there is enough sunlight to power the solar cells. Geologists attempt to figure out the hydrothermal dynamics of the crater, and biologists map the presence of microbes that grow beneath and even inside the rocks to avoid the wind, cold and fierce ultraviolet radiation of the Arctic, very much as life on Mars might do.

The ATVs can only carry around a small amount of science gear, and even when pulling a small trailer are limited for the most part to no more than day trips. A more substantial vehicle is needed to carry people, camping gear, and equipment around a larger territory, simulating how work beyond line-of-sight with camp might be done off-planet. So Pascal has negotiated a $1 three-year lease for a reconditioned Humvee from

AM General, which acknowledged it as the first official Humvee rover by giving it the serial designation of MARS-1. Pascal convinced the U.S. Marines to fly it up to Resolute in June 2002, last year; the problem now is how to get the hefty orange vehicle, originally configured to be an ambulance and far too large to fit on a Twin Otter, across Wellington Channel to Devon.

You could helicopter it across, of course, but getting a chopper that large this far north is a monumentally expensive proposition. An international conglomerate searching for minerals would do it without thinking twice, but it's the kind of expense NASA won't even consider. You could also ship it across the channel for about $20,000, but even that's more than the HMP budget can afford without some serious head scratching, and wouldn't be possible until the end of this coming summer at the earliest. That leaves driving it across now, while the channel is frozen, which has the added benefit of demonstrating to NASA how powered vehicular exploration can proceed autonomously under conditions analogous to those on Mars. As Paul and I are confirming, however, all activities in the Arctic depend on the weather, which this year hasn't exactly been cooperating.

The original plan was that Pascal and John Schutt, the HMP camp manager, would drive the Humvee down Resolute's one main street and out onto the ice in the bay, turn left at the entrance, follow the ice around Dungeness Point, and then head straight east across the Wellington Channel for Devon. Paul and his brother, Joe Amarualik, would scout ahead on snowmobiles to pick a safe route ahead of the Humvee, and I would ride swing in the rear and document the trip. As opposed to the snowmobiles, the loaded Humvee weighs in around 8,800 pounds. Pascal and John would wear the bright orange Mustang survival suits favored by the Coast Guard, and take turns driving solo with the doors off, which might allow them a chance to survive any unexpected plunge through the ice and into the water. In theory the entire project was supposed to take a week, but it's seldom that things run according to plan either on Mars or in the Arctic.

The warm currents that have cleared out the ice in Barrow Strait, and almost melted out the route from underneath Paul and me this morning, have also taken out the ice in Lancaster Sound and the southern half of Wellington Channel, a condition clearly visible to us when we flew into Resolute from Ottawa three days earlier. Pascal and Paul had already devised a second plan more than a week ago, however, when looking at satellite photos of the open water. It had been an exceptionally mild winter with very little snow, so why not take the Humvee north across the island until we found smooth sea ice offshore, then drive across? It was a question to which the weather replied with a low-pressure system that dumped three feet of snow on Cornwallis.

Paul had met us at the airport with the alternative route mapped out along the high ridges of the island where the wind had blown off most of the snow. But it still would mean driving through some deep snow in the valleys that had to be crossed. Plan number two came to a grinding halt when we cranked up the Humvee after its long winter hibernation and went for a test drive into a snowbank, where we promptly foundered.

When we'd flown in, images of Humvees were ubiquitous on televisions around the world. We'd just invaded Iraq five weeks earlier, and many of the journalists embedded with the troops were sending back constant footage of the vehicles plowing their way through the desert sands. Snow, as we discovered, is not analogous to sand. We couldn't even get the Humvee out of town and onto the ice of the bay.

Over the course of the next day we confirmed that the Humvee is a brilliant vehicle under many conditions. It can climb steep hills without a whimper, carry enormous amounts of gear, and when properly equipped, snorkel through water five feet deep. What it cannot do, at least as currently equipped, is handle more than sixteen inches of snow, the height of its ground clearance. The vehicle's tires don't provide enough flotation to stay on top of the snow, and once the vehicle high-centers and its four wheels lose traction, the only option is to shovel your way out. Much to our disgust, even the four-wheel-drive Ford trucks around town did better than the Humvee. In a pickup you could at least gun your way out of a snowbank. Try that with a Humvee and you run the risk of transferring the ten thousand pounds of torque from each of the other three axles into a single shaft—which will promptly snap under the enormous pressure. There are no spare Humvee axles in the High Arctic.

As Pascal put it: "What we need is ground with no snow or water with ice on top. Instead, we have snow and no ice." So, while Pascal sat in the Resolute Co-op Hotel trying to wheedle a set of rubber tracks from Mattracks in Minnesota—rubber tracks being triangular units you bolt onto each axle—Paul and I are checking the southern coastal route firsthand. On our flight in, Pascal had snapped digital pictures that showed a narrow border of fast ice clinging stubbornly to the shore around Dungeness Point and then up the coast until it met sea ice, maybe five miles to the north. There's a slim chance that there's still a route by sea, if not by land.

Paul and I had left the hotel a little after nine this morning, gassed up the two snowmobiles, and then driven out onto the bay. The ice was easy going until we hit the transition zone from last year's ice to the stuff formed this year, known as first-year ice. Wherever the fast ice from previous years and the new ice meet, in this case at the mouth of the bay, it's chaos. Driving the pressure ridges is like wending your way through giant shards of blue crystal, as if the sky were a chandelier that had fallen and splintered across the frozen sea. The jagged ice-scape is exquisite, but the work exhausting. You wrestle the handlebars of the snowmobile as you twist through narrow slots in between the upthrust slabs, and buck over three-foot-high hummocks disguised by drifting snow. Sometimes you're going so slow that your speed doesn't register on the speedometers.

Once through and onto the first-year ice, however, which is saltier and therefore more elastic and smoother, you can make great time, assuming it's thick enough. For the Humvee this means at least a foot or so with no nearby open leads, and that is not the situation this year at the southern end of the island. This does not bode well for the Humvee.

It's almost noon by the time Paul and I weave our way back through the pressure ridges and find the older multi-year ice near the shore. It's once again smooth riding, but we can see the open sea relentlessly encroaching, and the ridges being forced toward land, closing out any hope that the Humvee can make it via sea ice to Dungeness Point. We decide that we're going to have to climb up the steep hills of the shoreline to see if there's a snow-free route above us, but first pull off into the mouth of a steep-sided and narrow canyon to take a break.

Paul climbs stiffly off his machine and I do the same, both of us pushing down our hoods so we can talk. Paul's bareheaded. I'm wearing a black windproof hat and mask assemblage, and prying apart its Velcro strips sounds as if I'm ripping off my face. Once free of the dense fabric, I realize how stiff the wind has gotten. We're in the shade and the canyon, instead of providing shelter, funnels the wind with increased force. Snow, ice, bare rock. We won't stay here long. Paul crouches by the track of his vehicle and offers me some hot coffee from a thermos, which I happily accept. In return, I give him some of my peanut-butter-and-cheese-cracker snacks. One of the keys to keeping warm is a constant intake of water, carbohydrates, and sugar. When we finish the crackers, I root around in the top pocket of my pack for a chocolate bar, and consider the migration of humans over ice.

Mainstream scientific theories have it that people—either Homo sapiens or Neanderthals—have been in the Siberian Arctic for at least thirty-five thousand years, and perhaps as long as forty-five thousand. Whoever they were, they were not only sophisticated enough to carve tools out of bone, but also sculptures of animals, and to practice ceremonial burials. They didn't have refined sugar products with which to stay warm, but followed large mammals as fluctuations in the climate allowed both prey and hunter to spread northward. There was an ice age starting twenty thousand years ago that lasted for seven thousand years, when ice covered a third of the earth's landmass, but by 15,000 BC major deglaciation was beginning in North America.

The sequences of migration across the Bering Strait are multiple and to some extent still unclear, but it seems there were people in the Yukon living in caves at least as early as 13,000 BC and spreading southward. Those cultures inhabited the north of the continent up to the limits of the boreal forest, forcing the subsequent Paleo-Eskimo—Old Eskimos—to spread farther eastward from Alaska along its barren North Slope and across the Canadian High Arctic. As they migrated across the top of the North American continent forty-five hundred years ago, they walked the beach that Paul and I are sitting on. Ever since the Ice Age retreated, the islands have been rebounding in fits and starts from the enormous weight of the glaciers, slowly lifting up a sequence of firm gravel beaches that were both excellent platforms for settlement and pathways for animals and people.

The first people that we know of to migrate eastward through the High Arctic, starting in roughly 2500 BC, belonged to the Independence culture (named for the fjord in Greenland where the initial evidence was found of their occupation). They were musk ox hunters from Siberia who made microblades out of flint that they mounted into spears. They apparently lived in tents year round, and kept warm during the winter with tiny fires, burning moss and twigs. It's possible that they may have found the last remaining pockets of woolly mammoths, which had previously gone extinct in Siberia, still living on islands in the archipelago. When a significant cooling in the climate occurred around 1000 BC, land mammal populations decreased dramatically throughout the Arctic, and the Independence culture was succeeded by the Dorset people, who hunted small whales and seals from shore and sea ice. Their sustenance revolved around sea-based mammals, which do better in the cold.

According to the archeological record, the Dorset culture was apparently a peaceful one with a very highly developed aesthetic; they worked quartz and jade into exquisite tools and sculptures, and built boulder-and-turf houses heated with soapstone lamps that burned oil. They used bone and antler knives for cutting snow, and may have invented the domed snow house, or igloo. The Dorset were a very conservative people, maintaining their culture from Alaska to Greenland throughout most of their three-thousand-year history. Some scientists speculate that this homogeneity may have been genetic, and may have contributed to their demise when the next wave of immigrants arrived.

Paul's earliest ancestors, the ancestral Inuit known as the Thule culture, spread east from Alaska into the High Arctic within a single generation, and appear to have co-inhabited with the Dorset in some places from about 1000 to 1500 AD. But the Thule had brought with them a more adaptable technology that included large dogsleds, the recurved bows of the Mongolians, and a kayak-based hunting method that could take on the largest surviving mammals of the Arctic, the bowhead whale. We can trace how the Thule absorbed some of the Dorset skills and culture—such as the igloo—but there's little evidence that the Dorset learned from the Thule. And while the homogeneity of the Dorset people had let them trade among themselves clear across 150° of longitude, it may have made them peculiarly susceptible to Asian influenzas carried by the Thule.

The beaches of Devon, Cornwallis, and the other nearby islands host evidence of them all, stone from the oldest dwellings on the higher beaches, evidence of newer ones lower down as the islands rose. Thirty feet above the high tide mark are stone rings that held down the tents of the Independence people; a few feet down are hearths built of upright slabs from the Dorset culture; still lower on the slowly rebounding shoreline are remains of the circular stone and whalebone structures erected by the Thule people. A large encampment of a dozen Thule houses has been partially reconstructed outside Resolute, almost at the end of the runway where the remnants of a crashed jetliner sit. The shredded aluminum bulkheads and engine casings make an apocalyptic contrast with the stone and whalebone habitats of the Thule, and are a reminder that we won't be permanent residents here, either.

The Thule and their offspring, the modern Inuit, abandoned many of the farther reaches of the High Arctic during the Little Ice Age, which lasted from three hundred to one hundred years ago and obliterated the Thule whaling industry. Although an airfield and joint U.S.-Canadian weather station were built in Resolute Bay during 1947, Cornwallis wasn't really inhabited again until 1953, when the Canadian government decided to resettle some Inuit families from the Barren Lands to the south. The Cold War was on, and the government in Ottawa was nervous about the United States challenging its sovereignty over the Northwest Passage, a route that military icebreakers could use. The Americans had already begun to design the radar stations of the Distant Early Warning (DEW) Line that would run for three thousand miles above the Arctic Circle, and that they would begin to install the next year. Fifteen of the stations were to be placed in what was then the Northwest Territories (and is now the Inuit-governed province of Nunavut, which extends north from Manitoba to the Arctic Ocean). Possession is nine-tenths of the law, as the saying goes, and nothing signifies ownership better than inhabitation.

Despite the name, the Barren Lands consist of rolling tundra cut through by rivers. The countryside is rich in fish and caribou, and the Inuit had developed a superb land-based hunting culture. Resolute and Grise Fjord, a small town at the southern end of the much larger Ellesmere Island to the north, were communities hastily erected in an environment where the food source was primarily maritime in nature. The government had blithely assumed that any one place in the Arctic was as bountiful as any other for the Inuit, and dropped off only minimal supplies that the newcomers would supposedly supplement by living off the land.

The Dorset and Thule people might have been able to survive off the land and sea, but some of the Inuit in Grise Fjord starved to death that first winter. Ever since, the communities have relied heavily on a regular supply of food and goods flown and shipped from the south. People do hunt and fish, but without a major logistical pipeline from the south, they would perish; it's that simple. Mistrust of the government in Ottawa remains deeply embedded in the two communities, especially in Grise Fjord.

The sovereignty issues only become more complicated over time. The Arctic ice has been thinning by 3 percent annually since the 1970s, and any waterway that connects two oceans is considered to be international waters. The Canadians are not keen about seeing their northern border—and a key component of their national identity—besmirched by costly oil spills. The experience that Paul and I had earlier this morning gives proof to the pun that the Canadians are standing on thin ice. And then there's the issue of Nunavut, which was created as an Inuit-governed territory in 1999. Its 750,000 square miles hold only twenty-six communities with a total of fewer than twenty-nine thousand people, 85 percent of whom are Inuit. As a regional entity through which runs the heart of the passage, Nunavut lacks the resources to watch over the waters. Last summer the Canadian Navy sent a warship north for the first time since the Cold War, and both the Canadian and American navies are considering the need for a new generation of nuclear-powered icebreakers.

I look at Paul, who is sipping his coffee with bare hands. I've taken off my thickly insulated mittens but am still using my liner gloves, and my fingers, even around the hot cup, are getting cold. The wind chill has dropped again. "What do you think the temperature is?" asks Paul. I shrug. "Maybe minus twenty," I reply.

"It's okay when it's minus twenty. You get cold, but you don't freeze," he observes. I spent almost three months the previous year working in the Antarctic, where it got down to fifty below when I was at the South Pole, and I know he's right. But it doesn't make me feel any warmer. I hand back the cup and start rewrapping the face mask. Paul puts everything away, then comes over to adjust my hood, check my mittens, and in general satisfy himself that he won't be dragging back a frozen lump later in the day.

Paul was born on board the ship bearing his family north to Resolute, and counts himself the first native-born citizen of the hamlet. He's grown up on Cornwallis, been the deputy mayor of the town, and spends most summers out hunting or prospecting for diamonds on Devon Island. He's smart, funny, and knows what he's doing, but he's no mythical superman of the North. He can get into trouble in the cold or out on the ice like anyone else. It's just that he does a good job of lowering the odds, which I sincerely appreciate.

An hour later we're standing carefully to one side of a cornice on the shoreline ridge, the snow hanging over a cliff a hundred feet high. The pressure ridges below us are crushed directly into the shore, a narrow swath of tortured ice along which no vehicle can drive. Less than a hundred yards offshore a herd of walrus cavort in the whitecaps, led by a bull that must weigh fifteen hundred pounds. Walrus, unlike seals that swim using their hind flippers, oar through the water using their fore flippers. From our elevated vantage point, the half-dozen mammals below look like they're doing a powerful breaststroke as they rise and fall in the water, their long tusks gleaming. They dive periodically for the shallow bottom, where they're grazing on clams and other shellfish, using those tusks to stir the invertebrates up out of the silts. Paul points to the bull, who seems to have rolled over on his side and paused in his feeding: "I think he's watching us." He'd have to have good vision to do so. I'm following his movements through a small pair of binoculars and can't even see his eyes.

Across Barrow Strait are the high cliffs of Somerset Island, which slowly climb and then collapse as differential layers of temperature build and disperse over the water. The "inferior" mirages in hot deserts—where it looks like a pool of water is floating in front of you—are caused by a ground-level layer of very hot air abutting a cooler layer above. Where the two meet they create a mirror line that bends light, reflecting the blue sky toward you. The "superior" mirages formed over water occur when the thermal discontinuity is reversed—warm air over cold water—and what's reflected toward you is a magnified image of what's below or behind the mirror line, most commonly a far shoreline. The mirages can look like towering cliffs or castles, and even entire cities or mountain ranges. European sailors called them "Fata Morgana" after that archetypal femme fatale, Morgan le Fay, King Arthur's half sister and enchantress. They believed she lived in an undersea castle, and that she cast its image above the water as a lure for the unwary. Sir John Ross, sailing into Lancaster Sound on August 19, 1818, saw what he believed was an impenetrable range of peaks, and turned away. The following year Lieutenant William Edward Parry sailed to the same spot, saw no such landmass, and proceeded westward to Melville Island and almost to the far end of the archipelago, where his two vessels wintered over. Ross had been bluffed by a superior mirage.

The mirage of Somerset is seductive, but it's to the left that my eyes are drawn most insistently, even more so than by the presence of the walrus. Across the waters of Wellington Channel, so blue as to be almost black, Devon Island, and that spit of land connected to its southwest corner known as Beechey Island, are clearly visible in the hyper-arid crystalline clarity of the Arctic. The only permanent human residents of the island are buried there, the mummified corpses of three sailors from Sir John Franklin's expedition. They were the first of his crew to perish.

The geology and harsh environment of Devon Island make it a compelling physical analog for Mars, but this precise bit of history lifts it into metaphor. Franklin was an experienced Arctic explorer, and he assembled 128 sailors and officers to sail aboard the *Erebus* and *Terror,* ships that had explored the Antarctic coastline under John Ross's nephew, James Ross, two years earlier. The vessels, originally stout bomb ships built to withstand the recoil of heavy cannon fire, had since been reinforced and retrofitted with auxiliary steam engines, and were now state-of-the-art ice vessels. Franklin took with him more than 7,000 pounds of tobacco, 3,600 pounds of soap, 1,200 books on each ship, and enough food for three years, including tens of thousands of pounds of preserved meat, stored in 8,000 tins. His expedition was the most well equipped and thoroughly prepared in the history of polar exploration. Franklin was last seen in Baffin Bay headed for Lancaster Sound; he and his crew were never heard from again. In 1847 and 1848 the first three relief expeditions were sent to find him; by the time the search ended in 1859, a total of forty separate search parties had scoured the Arctic for traces of the missing men.

Searchers in 1850 found the first graves from the expedition on Beechey Island, three sailors who had died in 1846 after the party had wintered onshore. Franklin perished the next year, but based on evidence found to the south, the majority of his crew had abandoned ship and survived until a horrendous trek in 1848. What the searchers uncovered was very puzzling—the crew had walked away from where they might have met up with rescuers, carrying with them various luxury items, but leaving behind the food. Evidence of cannibalism was found in the remains.

The bodies on Beechey were exhumed and cursorily examined when they were first discovered, but it wasn't until the 1980s that a Canadian forensic team again exposed the bodies. Conducting an autopsy on the mummified remains of Petty Officer John Torrington, they determined that a contributing factor to the men's demise was lead poisoning from the soldering of the tinned food. In addition to suffering from scurvy, cold, and exhaustion, the sailors' cognitive faculties had been severely impaired by the toxic food. What had been an innovation in Franklin's time for preserving provisions had, instead, doomed the expedition.

NASA is on Devon Island to learn, among other things, how to avoid problems with its technology when and if it ever sends people to Mars. It's the agency's fervent prayer that it prevent not only explosive calamities like those that befell the *Challenger* in 1986 and the *Columbia* in 2003, but also potentially fatal mistakes in the design of pressure suits, habitats, radios, and other equipment. Getting to Mars, even with only unmanned probes, has a very poor record so far. As of 2003, NASA had only about a 60 percent success rate, and eleven of the sixteen Russian attempts had failed.

All of the Mars losses to date are mostly hardware issues. A human mission will also involve psychological risks. How do you maintain the sanity of a crew during a voyage that can take as long as nine months each way and would involve living together in intensely cramped quarters on both flights and on Mars? Well, you study how submarine crews fare during long underwater cruises, how people handle the six-month winter at the South Pole—and how the Inuit and Arctic explorers adapt to some of the most extreme conditions on Earth.

I snap out of my musings when Paul points to a low saddle between two small peaks nearby. "Let's go look up there. Maybe we can find a route the Humvee can use." We saddle up and pick our way carefully from snow patch to snow patch, trying to avoid rocks that will disable our rides. Within five minutes we're at the top and looking down into a deep gully. It's filled with snow. "This is too steep for us to drive down, even with the snowmobiles," Paul shouts over the engines. And just like that our reconnaissance forward is over. The only hope now is that Pascal can obtain the rubber tracks, and that they will float the boxy vehicle on the snow while driving north over Cornwallis to where the sea ice is smooth.

I find myself stamping my feet and waving my arms around in an effort to keep warm. My huge red down parka is the same model worn by all of us who have worked in the Antarctic, and it's commodious enough to curl up in and sleep, should your vehicle break and you have to wait for help. The insulated bibs and boots covering my lower half are what workers on Alaska's North Slope wear. The clothing is fine, but the wind has kicked up to nearly fifty miles per hour, and the ground blizzard has wrapped around this part of the island. It's time to cinch down the hood and drive the twenty miles back to Resolute.

CHAPTER TWO
Envisioning Mars

PAUL CIRCLES us back down to the shoreline and we plunge once again into the pressure ridges. The ground blizzard is also blowing down here and he pulls ahead of me—then just fades into the white wind. I don't know how tired I am until, heaving the machine around a block of ice the size of a grand piano, it bucks me off. I'm only going five miles per hour, and I watch myself fall in slow motion. Once my right hand lets go of the throttle, the machine immediately goes into idle and stops. My right leg is twisted awkwardly, the foot caught between the body and the faring. If I'd been going faster it would have meant a dislocated or broken bone. As it is, I struggle just to reach the handlebar and pull myself back on the seat. My right knee is sore, but my first concern is to get Paul within sight again. I rev up and push as quickly as possible along his tracks, which are beginning to disappear under the blowing snow. I find him stopped only a hundred yards ahead, waiting for me. He yells to ask if I'm okay; I just wave him on, disgusted with myself. The unforgiving environment here magnifies the consequences of every mishap, and people die in the Arctic from what elsewhere would be only a small mistake or mere lapse in attention.

The pressure ridges soon become impassable, and we edge back up onto the hilly shore. The wind is worse up here, but there's less ice in the air. I'm still warm from the exertion below, and when Paul unexpectedly peels off to one side and stops, I'm glad for a break. He's spotted a pile of stones the size of basketballs, and we dismount to investigate.

"Someone's been here. Look at this," he says, picking up a piece of rusted metal. It's a spring trap containing the lower leg bone of an Arctic fox. "Chewed off," Paul observes. Arctic foxes are solitary, forage widely over the land, and cache food for the winter. The fox would not have survived long on only three legs—it wouldn't have been able to run fast enough to catch the lemmings that are the staple of its diet. Seeing the tattered bone with fur still clinging to it underscores the soreness in my knee, the potential gravity of an injured limb. We push on to the top of the hill and the plateau, the wind blasting from our right and sending spindrift out over the sea. The entire landscape looks like it is running away beneath us. In a few minutes a small hut appears and we park in its lee, pushing with difficulty around to the side where an open door admits us into a dim interior. All that's inside are a couple of folding chairs and some brittle cables.

"What's this?" I ask, picking up one of the thin strands.

"Old listening posts for the NDE—the National Defense Establishment. They strung hydrophones from here out into the ocean to listen for whales. Or maybe submarines." Paul shrugs and we open up some salami and cheese that I took from the supplies in the back of the Humvee. "We're just above Assistance Bay where Sir Edward Belcher came through in 1852 looking for Franklin. He left food caches here for them." He pauses, looks at me. "We've found Thule skulls here. With bullet holes in them." Another pause. "I wonder if he shot those people to keep the caches safe." I have no way of knowing if what

Paul hypothesizes is true or not, but clearly he knows how colonial power works, whether it's the Royal Navy in the 1850s or the Canadian Navy more than a hundred years later.

Belcher's vessel, the HMS *Resolute,* was later nipped in the ice for two winters and he abandoned it. When the ship floated free, an American whaler salvaged it, towed it back to Connecticut where it was completely refitted, and then, politely and with more than a little irony, he handed it back to the British Navy. The ship was finally decommissioned and broken up in 1879, whereupon its oak timbers were used to make a massive desk that Queen Victoria presented to the U.S. president, Rutherford B. Hayes. It still sits in the Oval Office of the White House where it is used daily, a mute symbol of the connection between exploration, politics, and power. Paul knows the story well; I wonder if our current president does.

The final stop we make on the way back is at the half-buried wreck of a twin turboprop passenger plane that sits forlornly on the white plateau, nose pitched forward in the snow. Its tail is feathered in rime and the fuselage tilts gently to starboard. Great Northern Airways had high hopes for this plane in 1968, but when it ran out of fuel on its maiden voyage just a few miles short of Resolute, the company was unable to recover financially. A small mistake, big consequences. Now teenagers drive up here on snowmobiles in the winter and ATVs in the summer to scratch their names in the blue paint of the lower fuselage. Around us stretches a classic Arctic view, a highly isotropic snowscape. The seemingly unbounded space of gently rolling terrain is still blurred by spindrift. Paul looks around us, then at me. "Out here, you navigate by watching the sun and your shadow, so you know what direction you're going in and how to keep on course. But when clouds come, you have to remember what direction the wind was coming from, and use that. And you have to look at how the features in the snow have been made by the wind, so even if the wind changes, you still know where you're going."

He gives this lesson in navigation in the same spirit that he shares his coffee, because that's what people do in all desert environments. You share with the person next to you because at some point, you'll be depending on someone else for your survival.

By the time we're back in Resolute it's almost dinnertime, and we convene in Pascal's hotel room to debrief. Pascal, a planetary geologist who works at NASA's Ames Research Center near Palo Alto in California, was born in Hong Kong to a Chinese father and French mother, but was educated at a boarding school in Paris. He's a lean five foot ten, has short black hair, olive skin, and a perpetual grin that at the moment lurks within a sparse beard. When he's at home in Santa Clara he drives a black Corvette in between his two offices, one at Ames and the other at the nonprofit SETI Institute, the primary NASA grantee for conducting the "Search for Extraterrestrial Intelligence."

Pascal is addicted to challenges and is the proverbial perpetual optimist, which adds up to the perseverance not only necessary to convince Mattracks to donate part of the price of the tracks for the Humvee, but also for NASA to pay for the rest. He's beginning to accept that he may not personally walk on Mars—but he also understands that his tenacity will make it possible for others to do so.

We're joined by John Schutt, who mostly sits quietly in a corner. The fifty-year-old geologist, who is compact of build and wears his graying hair in a ponytail, spends the summers of the southern hemisphere on the blue glacial ice of the East Antarctic ice cap. He runs the camp from which scientists harvest meteorites found on the ice, and it's where he met Pascal in the mid-1990s. During the northern hemispheric summers he supervises the camp on Devon. John has spent so much tent time in the perpetual daylight of polar regions that he has to sleep with the curtains open when we share a hotel room. He's very good at being quiet and letting things happen; it's just that, around him, what happens goes well more often than not by a wide margin. That's an extremely desirable attribute when working in polar environments, places where almost nothing goes as planned.

Paul's younger brother, Joe Amarualik, is the other member of the team. He is the assistant HMP camp manager, and while Paul tends to be somewhat laconic, Joe is downright intimidating when you first meet him, he's so quiet. Last summer when I arrived on Devon, I think it must have been almost a week before I heard him say anything, and then it was four words murmured at such a low level that it took me another day to convince myself that he had, indeed, spoken. What I most like doing with Joe is watching him watch. Like Paul, his vision in the landscape is acute, and I've seen him spot a fossil on the ground twenty yards away while passing it at thirty miles per hour on an ATV. He once snowmobiled to the North Pole on a sixty-day trek.

Thus assembled, Paul and I give our report: that the route south is virtually impassable for a snowmobile, and certainly so for the Humvee. Paul again broaches the idea of heading inland north and then over the ice, this time on tracks, a plan that Joe thinks is feasible, having just returned from a trip to the far end of the island. Pascal, in turn, reports that he's been successful in actually getting the tracks. Unfortunately, it will take at least a week to ten days to get them to Resolute, make the drive to Devon, and return. He'll spend some of the time flying north to Grise Fjord on Ellesmere Island to work on getting permission for us to go inside Haughton Crater this summer, which is on IOL— Inuit Owned Land. John will wait here in Resolute, but I have a schedule to keep, so I'll head south to Ottawa to research artistic and cartographic images of the Arctic, then get back to the States.

At ten that evening I walk out of town and up into the hills until I can look over a ridge to the next valley. The snow underfoot crunches. The wind, which has never really stopped since we've been here, draws thin veils of spindrift off the heights, winds them around my ankles, then whips them down to the bay and out over the ice. More rocks are exposed tonight than last. During the lengthening hours of springtime daylight the snow is sublimating directly from the ground to the atmosphere without ever turning to water. Cornwallis Island is uncovered at this time of year by the dry wind, not melting, and if conditions stay like this, Pascal and John still stand a good chance of getting across the island to decent sea ice.

The sun is at my back and low over Resolute. The hills in front of me glow amber in the dim sunlight, but where shadows fall the undertones of the snow are blue. The landscape is pillowed and as empty in appearance as the hills and valleys on the moon that the Apollo astronauts photographed while walking on its surface. Inside my layer of windproof hat and parka hood, all I can hear is my breath. I turn and look at Resolute, its contours almost eclipsed by the needle-fine grains of spindrift. I wonder how the village will fare if the Northwest Passage opens and tourist ships start plying the waters, which at current melt rates could happen in less than twenty years. The only recent industry on Cornwallis Island has been the Polaris lead and zinc mine nearby, which after two decades has now petered out. The two Canadian airlines serving Resolute are cutting back their service, and with that constriction of the logistical pipeline supporting the community, it will become that much more isolated. It's easy to imagine the town disappearing forever under snowdrifts.

On the other hand, should tourism boom, commerce will follow—but most likely so will alcoholism, drug running, and crime, and the inevitable resentment toward the great multinational powers as consumerism increasingly displaces the traditional Inuit values that are based on solitary hunting. Then there are the diamond mines. Canada is now the world's third largest producer of the gems, many of which come from the Arctic. I half smile, thinking of Paul out hunting two months from now for a pipeline of the stones on Devon Island; the ones found elsewhere in the North during the last decade are projected to yield billions of dollars in revenue. Were he to be successful, mining could have an even more profound impact on this region than shipping or tourism, though that's no solution. Resource extraction, no matter if colonial powers or indigenous people run it, always carries its own steep environmental and social costs.

The contrast between the early hunting cultures and capitalism couldn't be sharper. The Independence people who wintered in tents four thousand years ago had only their occasional twig fires to thaw out slivers of frozen meat for sustenance. Winter darkness at this latitude lasts three months. Anthropologists speculate that they may have spent most of the time dreaming, whether awake or asleep. The Dorset people during the winter months produced figurines that still spook us, grimacing faces caught in the process of shape-shifting from human to wolf or bear and back again. These Neolithic cultures pushed human ingenuity and technology to the edge of the known world and well beyond it, surviving and at times even flourishing in both body and mind.

The passage to Mars is a dark one through interplanetary space, at least four to six months long and filled with the invisible hazards of solar storms and cosmic radiation. I cannot help but think the promise of resource extraction and commerce will lure people to force that route for trade, no matter the risk, just as the profits from crude oil beckon ships through the Northwest Passage. If the entwined issues of commerce and sovereignty are complicated in Nunavut, they will be positively Byzantine between planets, a situation that numerous science-fiction writers have used as the premise for novels about the colonization of both the moon and Mars.

I turn back to the hills with their ethereal glow, and wonder what dreams people will fashion when they get to Mars. I'll come back in ten weeks to hop a Twin Otter over to Devon Island, and will pass just north of the gravel strand on Beechey where the three sailors from Franklin's expedition are buried. They, too, had dreams about the glory of exploration and the grand profits of history. I'll keep that in mind when I land at the Haughton-Mars camp in three months to rejoin Pascal, John, and Joe.

How we conduct ourselves as a species in the polar desert of Nunavut—from the personal level all the way up to the geopolitical one—gives indications of how we will treat the empty reaches of Mars, a planetary desert equal in area to all of the combined landmasses on Earth. I ruminate over the intense and acute intimacy that Paul and Joe have with this landscape when traveling through it, and Pascal's perseverance, and how John refuses to let a piece of equipment stay out of commission even under the most trying of circumstances. I think about them driving to Devon, driving to Mars, driving to get to an edge of existence and endurance that the Paleo-Eskimo cultures would have understood.

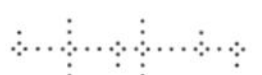

Everyone who lands on Devon Island to practice Mars arrives by a different trajectory, but more people interested in exploration seem to come out of geology than any other scientific discipline. That's true of both Pascal and John Schutt, both of whom I interview the next day while in Resolute waiting for my plane to fly out.

Actually, it might be more accurate to state that people interested in exploration tend to end up in geology, as it's an obvious way to combine a love of the outdoors with science—and a credible excuse to give one's parents for graduate school. In any case, it's where Pascal tells me he started, by earning a degree in geology from the University of Paris in 1987. He got his first taste of a polar environment the next year when he wrangled a spot as a geophysicist in the Antarctic at France's Dumont d'Urville Station, where he wintered over. He went not as much for the work as to insert himself into an environment on Earth that was as close as he could get to being on Mars, and he defines his life as "before and after d'Urville." For the first time he wasn't outdoors just for an adventure on a holiday, but actually exploring places where no one else had walked before. Having realized the addictive thrill of discovery, but knowing that almost everywhere on Earth has now seen human footprints is a powerful motive for wanting to experience firsthand the terrain of another planet.

A mecca for young scientists interested in the geology of other planets is Cornell University in Ithaca, New York. It was Carl Sagan's home institution, and today is where Steve Squyres, the scientific principal investigator for the Mars Exploration Rover Project, is based. After Antarctica, Pascal went to Cornell specifically to study with Joseph Veverka, an astronomer who had been a member of virtually every planetary imaging team from the mid-1960s through the mid-1980s, including the Viking and Mariner missions to Mars. Pascal was adamant about working on Martian geology. Veverka pointed out that there would be no new data—hence no material for a PhD thesis—until the *Mars Observer* reached orbit around Mars with its camera in August 1993. But, two weeks after Pascal arrived, Veverka took him to the Jet Propulsion Laboratory (JPL) to sit in on the *Voyager 2* encounter with Neptune. Pascal's face lights up when he recalls the experience.

"Up until then, no one had seen Neptune close up, and we had no idea what it was like. I'd grown up during the Voyager missions—1979 Jupiter, 1981 Saturn, 1986 Uranus, and then 1989 Neptune. So there I was, this young guy just arrived at Cornell, and I was part of that. It was incredible.

"I like a planet with a hard surface where there's geology to be done, so I studied Triton, Neptune's largest moon and one of the largest in the solar system. I studied its impact craters, and learned how to be a planetary scientist in two years at Cornell, but my thesis was always supposed to be the Mars Observer mission.

"A year before the scheduled orbital insertion, I spent twelve months preparing the photo sequences for the 1.4-meter resolution camera. I helped my adviser select over one thousand sites, applying various criteria to select features to be photographed, as well as the filters, resolution levels, and so forth. It was my first in-depth task and I spent nights up working on it, using the low-res images from *Voyager* to choose the shots, imagining what this high-res camera was going to show."

Just before insertion, Veverka sent him to San Diego to work at Malin Space Science Systems, the company that built the camera and would process the data. "But I was only there for two weeks, because the insertion failed, the propellants for the retro rockets mixing in the wrong place and turning the orbiter into a technological asteroid, a T1. We have all these classes for asteroids, A, B, C and so on—I think we should have a new category for the ones we build!" Pascal shakes his head, thinking of all the spacecraft we've lost trying to reach Mars.

"So I went back to Cornell. *Galileo* was on its way to Jupiter, and it flew by the asteroid Gaspra, which was named after a spa in the Crimea near Yalta. Joe was tasked with naming the impact craters on it, but he gave that to me, so I looked up spa names from around the world from Baden-Baden to Calistoga, and named them."

Pascal was studying both an impactor—an asteroid—as well as impact craters. This turned out to be fortunate, given that he wanted an excuse to return to the Antarctic, since the only planetary science done there is looking for impactors. He went on to obtain both a master's degree and PhD from Cornell in astronomy and space sciences, finishing in 1997, but during the previous year he joined the U.S. Antarctic Search for Meteorites (ANSMET), which is where he met John Schutt. The team spends months camped in tents out on the hard blue ice surface of the glaciers up on the East Antarctic plateau, and methodically searches the ice in a grid pattern for meteorites. It was another member of that team, Robbie Score, who in 1984 had found an unusual specimen that was identified a decade later to be of Martian origin. Then in 1996 a group of scientists copublished a paper laying out four separate lines of evidence indicating that the rock contained fossil evidence of nanobacteria on the neighboring planet. It was a controversial claim and one that remains unsettled even today, although mainstream opinion is that what appeared to be organic fossils are more probably of geological origin. Nevertheless, ANSMET is the most productive meteorite search in the world, and another formative experience for the young scientist.

Pascal first became interested in Haughton Crater while he was still at Cornell, when he used to drive north from Ithaca to visit impact craters in Ontario, Canada. He would identify them from satellite photos, assess them further through examining aerial photographs, then go visit them. As often seems to be the case with people passionate about going into space, he's an avid pilot and became an FAA-certified flight instructor for helicopters in 1993. He had the idea that he could hopscotch in a helicopter from crater to crater until he ended up at Haughton, then the northernmost such feature known in the world. He had been in Resolute earlier that same year to help a friend organize a ski expedition to the Magnetic North Pole—not the terrestrial pole about which the planet revolves, but the ever-drifting convergence of geomagnetic lines, then located six hundred miles south in the middle of the frozen Arctic Ocean. When he was unable to convince a helicopter company to sponsor the adventure, he contacted Polar Shelf, the science logistical operation in Resolute, to see if they were interested in donating support for his flight over to Devon. They weren't, so getting to Haughton would have to wait until the next stage in his career.

When Pascal left Ithaca in 1997, he switched coasts to be a postdoctoral research associate at Ames. Once there he found an enthusiastic and visionary colleague in Chris McKay, another planetary scientist and one of the leading proponents at NASA for the human exploration of Mars. McKay gave him the support necessary to put together a small Haughton-Mars Project research team. That summer Pascal flew with Schutt and two other scientists into the crater to assess the campsite that had been the base for previous geologists on Devon. Its broad gravel terrace made a natural landing strip for the Twin Otters, it was near a source of water (the Haughton River, which drains the crater), and it offered a number of geological features that might be analogous to those on Mars.

The reconnaissance turned out to be incredibly successful. The frigid and remote island offered fossils and geothermal features that exposed the story of the Eocene impact that created the crater, and periglacial features left after the retreat of the last ice age included meltwater channels and gully systems perhaps similar to those on Mars. The view from the men's tent doors encompassed high cliffs, the river, and broad swaths of rock shattered by the impact—the breccia that gleamed like silver in the low light. It was stunning. As Pascal says of both Devon Island and the Dry Valleys in the Antarctic, "Each one is a little landscape of another world."

The next year twenty-four team members from NASA, and various universities and institutions in both the United States and Canada, visited the camp and conducted research in geology, biology, and remote-sensing, as well as exploration techniques, such as ground-penetrating radar, permafrost drilling, and robotic helicopter tests. The project was gaining enough attention that *National Geographic* sent a writer and a photographer to do a piece about it. Michael Long's article spelled out the science being done, and the photographs by Peter Essick remain the most haunting images of the crater's austere beauty.

That fall Robert Zubrin, a former nuclear engineer turned Mars enthusiast, showed up at NASA Ames to float the idea that his organization, the Mars Society, be allowed to put a small instrument aboard a NASA spacecraft. That project didn't pan out, but soon thereafter Pascal proposed that, since the goal of the nonprofit group was to put humans on Mars, the two organizations should work together on Devon. Because the camp in the crater was located on Inuit Owned Land, and the Inuit authorities were becoming increasingly reluctant to grant permits for the site, Pascal was thinking about building a permanent facility outside the IOL, somewhere on or near the crater rim. This way the Mars Society would have a project around which to raise funds, and the NASA scientists a place from which to conduct research. Zubrin went away saying he would think about it, and apparently the idea grew on him, because he accepted an invitation to visit Devon during the next field season.

In 1999 forty people flew into the crater, including scientists from Russia and the United Kingdom. At the end of the field season NASA formally established the Haughton-Mars Project with Pascal as the principal investigator. Zubrin, Pascal, and John Schutt visited the short list of possible sites for the new camp. Pascal had developed thirteen criteria for the new place, among them the proximity of water and a landing strip, but the place also needed to accommodate a construction camp that could be hidden from the sight of inhabitants working in a mock habitat to be built by the Mars Society. His first choice was high on the northwest rim of the crater where the surface was rocky and the view looked remarkably similar to the panoramic images that the Pathfinder mission had sent back two years earlier from the floodplain of Ares Vallis. Three objectives underlying the criteria were kept in mind. One, what would be a practical site for polar logistics, which aren't trivial. Two, what would provide easy access to good analog sites at which both to practice exploring Mars, and also to do science relevant to future Mars missions—the idea of practicing Mars on Earth. And three, what would look good in the media through a red filter on a camera lens.

The site on the crater rim was prone to be windy, which bothered Zubrin, but it had a fine view out over the crater, was close to a small stream that ran during the summer, and there was a flat spot nearby that was just large enough for an airstrip that could accommodate the Twin Otters flown out of Polar Shelf in Resolute, which could bring in more cargo at cheaper rates than the helicopters. By fall of that year the Mars Society had hired an architect and was fundraising for the Flashline Mars Arctic Research Station (FMARS).

The turn of the millennium saw the largest contingent ever to work in the crater, almost a hundred participants. A construction camp was established about three-quarters of a mile away, and included most of the tents in our current camp. The U.S. Marines, who were supporting the NASA project, agreed to paradrop in the prefabricated steel and fiberglass components of the two-story cylindrical habitat, the "hab," and despite a fifteen-thousand-pound palette destructing upon impact, the structure was completed by the end of the summer. While the hab was being built that July, Pascal's project conducted research. Geologists made a five-day traverse to the opposite side of the crater; a limnologist sampled lake waters and sediments for diatoms, and a biologist looked for endoliths, the bacteria that live inside rocks. Hamilton Sundstrand engineers flew in an early version of a planetary exploration concept space suit for field tests, and *Canadian Geographic* produced an hour-long television documentary on the project titled "Mars on Earth."

The 2001 field season was run out of the hab under the joint direction of Pascal and Zubrin. The HMP project was now being managed by the SETI Institute with Pascal as the project lead and principal investigator. Zubrin concentrated on promoting mission simulations, or "sims," where people wear mock-ups of pressure suits and use the airlock whenever they exit the hab to do EVAs, or extravehicular activities. The sims were good for public relations and fundraising, as well as for testing exploration constraints. The NASA/SETI Institute work was providing valuable research into both geology and biology pertinent to Mars, as well as testing its own set of exploration metrics. The FMARS was occupied by six crews, each comprising five to seven members living in the station for four to ten days.

In the fall of 2001, the Mars Society and NASA got divorced. It would be too simple to say that it was a clash between the personalities of Pascal Lee and Robert Zubrin, although both have strong egos. But Zubrin had swung the Mars Society around to place more emphasis on public relations and advocacy than science, versus the HMP's emphasis on collaborative science and exploration research. Pascal likes to say it's the difference between church and state, where the Mars society represents the former.

My trajectory to Devon, like Pascal's, was routed through the Antarctic. I first landed at the HMP airstrip in July 2002 on a Twin Otter alongside a half-dozen engineers, an experimental pressure suit, and a box of homemade brownies—but six months earlier I'd been in the Antarctic as a visiting writer with the National Science Foundation. For the last decade I'd been working on a series of books about how we enter a space—somewhere we haven't been before—and transform it into a place. Or how we turn terrain into territory, or land into landscape. The terms depend on the context, whether you're thinking about the process as part of the evolution of human neurophysiology, the geopolitics of exploration, or the history of culture. The evidence I examine includes nonfiction accounts of exploration, fictional stories, and myths, but mostly the traces left in visual culture, such as maps, paintings, photographs, and architecture.

Eighty percent of everything we learn daily is information we collect through vision, our predominant sense, so how and what we see as we alter terrain is critical to the process. The process is more easily observed in deserts than in more temperate climates for a number of reasons. Not only are there fewer trees, but the layers of culture are thinner. You can see more easily how we deploy technological and cultural means, such as maps and paintings, to augment our limited visual neurology as we try to navigate in a difficult environment. So I'd been traipsing across the desert of the American Southwest to examine how we spread a cartographic grid over its great arid distances, and how that had translated into the enormous urban grids of Los Angeles and Phoenix. And I'd been in the Antarctic to see what kind of artistic, cartographic, and scientific images people produce in the most hostile conditions on the planet.

A couple of months before leaving for the Antarctic, a friend asked me what was next: Mars? I grinned and shook my head, aware both that the timing was premature and that my temperament is hardly astronaut material. But I remembered that a group of NASA scientists was testing exploration methods in the Arctic, and I wondered if I could wrangle a visit with them. Just before flying south in late 2001, I met Pascal for lunch at a sushi restaurant in Santa Clara.

When I had first called Pascal to express an interest in visiting the Haughton-Mars Project, he queried me sharply about what I could contribute to its mission. As with most science and exploration parties in the deep field, Antarctic or Arctic, everyone is expected to work as a member of the team, as well as to perform their own investigation. I told him that I wanted to investigate the effects of cognitive dissonance in isotropic spaces as it applied to Mars—and that I was good at washing dishes. He was intrigued enough to invite me to meet, and we spent two hours over lunch discovering our mutual interests. By the end of the conversation we were discussing the specifics of how to book plane flights from Vancouver to Yellowknife, then on to Resolute and Devon.

While in the Antarctic for the next three months, I interviewed helicopter pilots, long-distance cross-country skiers, and scientists about cognitive disorientation in whiteouts—where everything looks the same in all directions—as isotropic a situation as you can find on Earth. I compared paintings and maps of the enormous spaces with what I saw while walking around the South Pole and through the spectacular Dry Valleys, a 1,370-square-mile region in the Transantarctic Mountains where it has not rained for more than two million years. From time to time I wondered what the differences would be in the Arctic. The Antarctic is a frozen continent with not even any land mammals, much less indigenous people. The Arctic is a circumpolar ocean surrounded by people who have hunted across its frozen surface for thousands of years.

While traipsing through the Dry Valleys, I was constantly in the presence of experiments related to the exploration of Mars, the valleys being one of the better analog environments on Earth for Mars. On frozen Lake Bonney, for example, graduate students were studying how algal mats could grow under a permanent cap of ice twelve feet thick, and other biologists were studying bacteria inside the rocks—the sorts of places where life might be expected to shelter from the extreme cold and radiation found on the unprotected Martian surface. NASA had test-driven rovers through the valleys, and the novelist Kim Stanley Robinson, who had also been a visiting writer there, began his legendary science fiction Mars trilogy with the astronauts training in the Dry Valleys.

When I met with Pascal, he had rattled off a list of "Mars on Earth" locations: the Dry Valleys, Iceland, volcanoes in Hawaii, the Sahara, the Atacama Desert in Peru, the Tibetan Plateau—and I had realized how many of them I'd already been to, or were on my list to visit. The convergence between cognition, deserts, and Mars was obvious, and I kept running across examples while on the polar continent. One night I walked out onto the incandescent blue ice of Lake Vanda, a permanently frozen lake in the Dry Valleys. It was patterned with geometrically arranged white cracks that created the illusion of my standing upside down on a great vaulted ceiling. I knew that the polygonal patterns found on the frozen floors of the Dry Valleys were the same as those found in the dry lake beds of the Great Basin in Nevada, but I had no idea that these same stress patterns occurred in the surfaces of the only permanently frozen lakes in the world—and were also the basis for the design of the geodesic dome at the South Pole. Both are described by the principle of tensegrity. In architectural terms, a tensegritic structure is one where compression and tension are evenly distributed and thus balance each other. The principle operates at both micro and macro levels in biology, governing the design of everything from cell structure to skeletons. Several days later, when writing up the parallels, it seemed only natural to discover that such patterned ground had also been photographed on both Devon and Mars, evidence that on the latter, ice near the surface has recently undergone some kind of thermic cycling.

Another revelation was the value of changing one's sense of scale when looking at the Antarctic. I had started with the premise that we were forced to rely mostly on our visual imagination to understand a place so geographically remote and shrouded by cyclonic storms that it wasn't until 1997 that it was finally charted from space—twenty-five years after Mars had been mapped. The continent remains the coldest, windiest, highest, driest, and most remote landmass on Earth, but I came to realize that it's no longer a terra incognita. Our minds can grasp it. One reason why is that, despite what appears at first to be a vast sterility, the Antarctic is in actuality shot through with life. It's mostly microbial, but it's there, buried even in the ice of the South Pole.

In order to discover and then absorb that fact, I had to change the cultural framework from within which I viewed the Antarctic. Most of us, having been inundated with stories about Robert Scott and Ernest Shackleton, imagine the continent as a blank and perilous wasteland to be conquered only with heroic efforts. The notion that life has adapted successfully to the continent demands that one obtain a more intimate view, from a humble posture, preferably on one's knees, nose pressed to the ground while looking for bacteria inside rocks. This, too, I realized, was applicable to Mars. The search for life on the Red Planet wasn't going to be about scouting its hillsides for alien cities, but a patient and very careful probing for microbes based on techniques developed, in part, on the glaciers and frozen lakes in the Antarctic.

I returned to the States in January 2002, worked on the Antarctic book for a few months, then flew to Resolute in July. From there I boarded a Twin Otter to fly across Lancaster Sound and into the polar desert of the world's largest uninhabited island. I spent two weeks at the Haughton-Mars Project on Devon that season, accompanying the scientists as they pried open rocks, waded into icy lakes, and attempted to perform fieldwork in the pressure suit while talking on the radio via a portable relay station to camp. When in camp, I wrote up my field notes, washed dishes, and interviewed people. If anything, Devon was an even better analog in my mind for Mars than the Dry Valleys had been. It wasn't as cold, nor was the scenery as dramatic as in the Antarctic, but its bleak rock-strewn plateau simply looked more like the images *Pathfinder* had sent back in 1997.

On the final morning of that first visit to Devon, I walked up to the rim of the great Haughton Impact Crater. The depression is a silvery expanse so broad that it contains entire ranges of hills and the robust Haughton River. It's almost impossible to assess it visually without the aid of a map and aerial photographs, or some kind of elevated view from which you can encompass enough of it to grasp the effects of the impact. I sat on the rounded edge of the eroded rim and thought about how we overlay deserts with a cartographic grid in order to measure ourselves within them, how we rule lines over a flat abstract picture as if drawing on the land itself—and how we translate those lines into canals and roads, seeking to overrule desert aridity. Sometimes we construct this geometry in reality, such as with irrigation canals in Arizona, but sometimes the patterns exist only in our imaginations.

The amateur astronomer Percival Lowell was a supreme example of how cognition can overstep reality. In 1877, during one of the closest approaches to Earth that Mars can make, the Milanese astronomer Giovanni Schiaparelli observed the contrasting bright and dark areas of the planet, naming the dark ones after the Mediterranean and North seas, the bright ones for terrestrial deserts in the Middle East. He also sketched what he took to be intersecting lines on Mars, labeling them *canali,* or channels. He thought they crossed the Martian continents from sea to sea, and declared that they were as precise as if they were lines drawn with a pen. Eleven years later, Schiaparelli wrote that other astronomers earlier in the century had also noted the presence of the lines. Lowell, son of a Boston Brahmin and sister of poet Amy Lowell, had graduated from Harvard in 1876 with a degree in mathematics, but abandoned the life of a wealthy aristocrat to live as an expatriate in Tokyo and write books about Asian culture and mysticism. Becoming fascinated with Mars, he returned to the United States to build a private observatory in 1894 atop a forested hill at seven thousand feet in Flagstaff, Arizona. For fifteen years he peered through the eyepiece of his twenty-four-inch refractor, over time claiming to have mapped more than five hundred "canals," as he chose to translate Schiaparelli's term. The year prior to the opening of Lowell's observatory, the Italian warned that the *canali* were more likely the result of natural processes on Mars than the work of intelligent beings, but Lowell was so persuasive that later in the decade even Schiaparelli came to believe that the geometry could not have simply evolved by chance.

Lowell published *Mars,* the first of his three books about the planet, in 1895. In it he insisted that the lines appeared "absolutely straight from one end to the other, or curved in a equally uniform manner," and that "their most instantly conspicuous characteristic is this hopeless lack of irregularity." He drew them as crossing at circular points, which he supposed to be oases, and stated that the network divided up the planet into a system of "spherical triangles." His maps bear more than a passing resemblance to the patterned ice and ground I'd seen in the Antarctic and on Devon. Despite the fact that a majority of the world's astronomers could not confirm his observations, even scientists using instruments of superior size—such as Edward E. Bernard at the Lick Observatory, and George Hale working with the sixty-inch reflector on Mount Wilson—Lowell's views were grasped by a public eager to project a collective fantasy upon a new shore.

One of the most persuasive arguments against Lowell's visual interpretation of the planet was made by the British astronomer Edward Maunder, who had been photographing and studying the cyclical behavior of sunspots since 1873, and who thus had extensive experience in eliminating false interpretations of visual phenomena. As early as 1894 Maunder conducted experiments in order to determine what the smallest dots were that could be detected with his eyes. To his surprise, small dots invisible as single entities became visible as a straight line when placed not immediately next to one another, but in proximity. He surmised that Lowell might be unconsciously aggregating details too fine to see individually into geometric structures. Lowell's assistant, A. E. Douglass, performed similar experiments in 1897, after Lowell claimed to have seen lines radiating like spokes from a point on Venus. Douglass set up artificial planetary disks and peered at them from a mile away; much to his discomfort he saw illusory markings like those claimed by Lowell on Venus. He was fired after communicating his dissatisfaction with the astronomer's "unscientific" methods in 1901 to Lowell's brother-in-law. (Douglass later went on to found the University of Arizona's observatory in Tucson.)

In 1903 Maunder set up an even more rigorous test, an elegant and early experiment in cognition in that it demonstrated again how human brains organize random dots into lines, a finding confirmed by scientists studying boundary recognition in visual systems later in the century. Maunder drew an assortment of dots on a disk, and asked students to observe the disk and then draw what they saw. The farther away from the disk they were, the less clear became the individual marks—and the more the students showed a propensity to invent lines, integrating them together into recognizable figures. Lowell, who lived from 1855 to 1916, had witnessed the construction of both the Erie and Suez canals. His mind did what our visual systems will do when confronted with ambiguous information, in this case the blurred telescopic view of a rapidly rotating planet: It took adjacent areas of differing shades and regularized the boundaries into straight lines. Then his mind conflated what he thought he saw with human engineering, and possibly the arid Arizona environment, and stitched together the fuzzy observations into a globe encircled by canals, supposedly built by an alien race in a desperate attempt to save itself through irrigation.

I resumed hiking along the shattered curving ridge of the crater, amazed at how difficult it was to establish a sense of direction when both the landscape and path of the sun curve around you. And I thought about Sir John Franklin's legendary expedition in search of the Northwest Passage—another kind of *canali*—and how multinational corporations were now eying its use as a commercially viable route for the first time in history, a development that worried both the Canadian government and the Inuit provincial government of Nunavut. Exploration in the Arctic was one thing, colonization another, and the Canadians had sent a warship north that month to remind everyone that they claimed sovereignty over the passage, a claim not honored unanimously by other countries, including the United States.

This echoed the split between many members of the Mars Society and the NASA scientists. Zubrin believes that it is human destiny to colonize the other planet; the NASA people hew more to the viewpoint that it would be a fine place to explore but a lousy place to live. History curves around too, and Devon Island remains rich with associations in the history and contemporary practice of exploration and the attendant politics, all of which are based directly on our relationships to the landscape itself. I envisioned a book that could, as in my Antarctic work, use cognition as a thread to follow in this dense tapestry. Once again I could rely upon my travels in the field with scientists and others as a narrative armature around which to wind that thread—but it would mean convincing Pascal that I needed to return to the island.

That last afternoon, while I was waiting to fly out on the Twin Otter, Pascal debriefed me about the work I'd done during the two weeks, something he does with all departing team members. I gave him my initial impressions about how I found the dendritic and confusing maze of meltwater channels on Devon Island to be the most complete analog for Mars that I'd seen, and how wearing the pressure suit compounded the cognitive difficulties that humans face in unfamiliar terrain. I mused that the experiences of polar explorers could illuminate the challenges to be faced on Mars, and that, given the high degree of science-fiction literacy among NASA personnel, I needed to look at the artwork, novels, and movies about the planet. Inevitably those images have created expectations for the conditions we might find there. He grinned as he took notes, nodding his head.

"So Bill, listen. We have this Humvee sitting in Resolute that we're hoping to have here next summer as a mobile platform for science experiments. We're going to get it here by driving it across the sea ice next spring, and I think you should come along—if you're interested. And if you want, you can leave your sleeping bag and tent here for next season."

I took a deep breath. Driving to Mars? Who could resist?

CHAPTER THREE
Roving

KEITH COWING'S ATV is in mud up to its axles. He's in so deep that he's having trouble depressing the foot-operated gear lever to put it into reverse so that Gordon "Oz" Ozinski can pull it out from his secure perch on a patch of stony ground. We're in the middle of a valley tucked along the western side of Haughton Crater, and Rhinoceros Creek braids all around us. Tiny yellow poppies are in bloom and red sorrel patches dot the hillsides. Every few yards Arctic willows no higher than the sole of my boots hug the ground. And the mud is endless, fed by the snow that's visibly melting back by yards every day during this sunny July. The contrast with the blowing snow in Resolute three months earlier is not, however, entirely unwelcome.

The ground of Devon Island sits atop a layer of permafrost that begins anywhere from eighteen to thirty-six inches beneath the surface and extends downward for more than fifteen hundred feet. As a result, the ground is unable to absorb much runoff or meltwater; it just gets saturated and waits for someone to try and walk, run, drive, or otherwise transit across its surface. The part of the ground that thaws out seasonally is known as the "active layer." The name is appropriate—the stuff seems to grasp your feet with the conscious intent of stripping off your boots, socks, pants, and long underwear. But at least it's warm enough that, should you lose a boot, you won't be frostbitten within seconds.

The Humvee team had, indeed, made it across Cornwallis and the sea ice in early May using the bolt-on rubber tracks. They left the vehicle parked on a bluff above Wellington Channel until a week ago, when Pascal, Joe, and two other crew members flew out by Twin Otter to drive it the sixty miles to our base camp at the crater. Including detours necessitated by the terrain, they had already driven that many miles in three days before being stopped by a river that, when it's colder here, would be a passable stream. Now it's a briskly flowing watercourse a hundred feet wide with channels cut into the gravel that are much deeper than the eighteen inches or so that the Humvee or ATVs can manage. Then there's the issue of the bolts that have shaken out of the track units, but which the company assures us are not absolutely essential.

The crew cleared a short runway by what they named Endurance River, and an Otter was able to land and bring them back to camp. They're due to go back in a couple of days—when most of the snow may be gone and the temperatures may have dropped—to see if the river has gone down sufficiently in the early morning hours to be crossed. If so, they'll be continuing along a route no one has ever traveled before. They'll still be facing the kind of mud that we're floundering in this Sunday afternoon, and the route will end up being twice as long as they had originally plotted.

Devon Island is the size of West Virginia, which makes it roughly twenty times larger than the Dry Valleys in the Antarctic. It runs from east to west for about half its length, then makes a dogleg northward. The crater sits within the upper half of the dogleg, closer to the north shore of the island than the south. The average temperatures here during the summer, which lasts from now in mid-July until the middle of next month, run from a high of 44°F to a low of 27°F. This afternoon, however, it's almost 50°F, and we're basking in the fifteenth day of clear sunny weather. Last year at this time it was mostly bitter sleet, snow, and wind. You didn't want to walk from your tent without wearing serious mountaineering gear, and everyone huddled around the propane heater in the kitchen tent. The upside of the unusually warm weather this year is that I'm dressed in only lightweight long underwear, jeans, and a pile jacket. The downside is the mud, which fountains up from the tires of the ATV as it finally lurches toward solid ground. Keith is mud-spattered from head to toe and looks like he's been running a road rally in Central America.

Oz, on the lead ATV of our four vehicles, is a twenty-seven-year-old Scotsman who has just finished his dissertation for a PhD from the University of New Brunswick. He earned the doctorate by coming here for four years and systematically mapping and sampling the entire crater, spending more time on its floor than anyone else in the world. Like most field geologists, he's handy with a shovel and a towrope, as is John Parnell, another geologist with us who is helping push the ATV out of the mud. John, a slender man in his midforties who wears glasses, directs the Geofluids Research Group at the University of Aberdeen in Scotland. It's his second season here collecting rocks that he then takes home to analyze for minute fluid inclusions. What he's looking for are hydrocarbons—those compounds formed by the decomposition of organic material—and other clues about the chemistry of ancient water-and-organics-rich environments before, during, and after the Haughton impact event. This is exactly the kind of work that needs to be done on the cratered surface of Mars in order to determine whether or not its salty seas contained life.

Devon, like the other islands nearby in the Queen Elizabeth group, is a relatively low-lying body of land with cliffs along parts of its seashore.

It's riven with meltwater streams and rivers that branch and branch and branch again, a maze through which there is no simple route, much less a defined path, except those carved near the crater by the ATVs over several seasons. It might rain as much as one inch in July, but precipitation annually totals less than five inches. Despite the seasonal mud, that scant amount defines Devon as a desert; our tracks can last for decades, so we stay within established ruts whenever possible. It's not that anyone else will ever see the marks we make here, but there's no one working on Devon who isn't aware of the responsibility of choosing to write our presence on the land. Unfortunately for Keith, who's bringing up the rear, the saturated ground is thixotropic: Shake it and it liquefies. The three of us ahead of him have done a good job of jostling the track and it's the fourth time he's gotten mired.

We're on our way, if haltingly, to a small feeder meltwater stream that cuts through the ancient lake bed sediments of the crater and runs into the creek. Yesterday Oz found the fossilized skull of an extinct giant hare, a Miocene species that lived here twenty-plus million years ago. In a reversal of life as we know it, the hares on Devon were the size of large dogs, but the woolly rhinoceros for which the creek is named weighed in more on the order of a pig. The island itself was lightly forested with birch, pine, larch, and spruce. The temperatures then would have ranged from a low of 5°F during the winter nights up to a high of 80°F at this time of year, and the growing season was between three to five months. In short, it was more like Toronto is today, and we would have been riding around in T-shirts. Oz is hoping to find more fossils, and the rest of us are tagging along for a variety of other purposes.

Once Keith is moving again, Oz rides across the creek to my side. John splashes across on foot to where he'd left his own ATV. The water is only ankle deep, less of a problem than the mud, which can suck you in up to your knees. John's research group studies the origin, migration, and evolution of fluids in Earth's crust, and he does not look at all happy, reflecting the fact that, although he is an avid hiker, he'd prefer to look at water inside rocks through a microscope than wade through water flowing over them. He's been quite pleased, however, when breaking rocks open in the crater to smell the telltale sign of ancient organic material that we associate with petrochemicals. Inclusions of water in rock on Mars would be a stunning discovery, and he's constantly looking at the rocks on Devon to assess the likelihood.

John, when he speaks, sounds very much the proper Londoner he once was. Oz comes across as the rock climber from Scotland that he still is, a hint of burr and roll in his accent. Keith, a former rock climber, has a trace of what I think might be New Jersey in his accent, and is the most irreverent person in camp. Although he's here to manage the ongoing construction of the Arthur C. Clarke Greenhouse (named for the science fiction author of *2001: A Space Odyssey*), he makes what he refers to as "a dot-com living." He runs www.SpaceRef.com, the definitive online reference source for space exploration, and the much rowdier www.NASAWatch.com. He's pro-space, but a sharp critic of the space agency. Keith once worked in the space station program for NASA, and before that in its life sciences division—biology in space. His stream of verbal humor is sharp and nonending, and it's easy to see why he and a government bureaucracy weren't a perfect marriage. He handles mud with equanimity, perhaps a prerequisite for working in greenhouses. He's along for the ride, as am I, to write about it, as well as to take a break from staring at a computer screen all day.

The four of us continue upstream past the twenty-million-year-old mud hills, careful to stay together as a group, not just to provide towing services as needed, but to keep anyone from getting lost, which used to happen here before team travel was made mandatory. One reason that this part of the Arctic archipelago is so confusing to human beings is that—with the exception of the impact—it hasn't been a very dynamic landscape for the last four hundred million years. The sediments on Devon date from the Lower Ordovician through the Lower Silurian, about 480–400 million years ago, when these islands were surrounded by a shallow tropical sea. The brown limestones and dolomites around our camp were laid down as the slow drift of unicellular sea life came to rest on the ocean bottom. Imagine the Caribbean today and you get the picture. People are used to living in environments that are much more dynamic and show greater effects of erosion. As a result, Devon appears to us as a somewhat unearthly terrain.

The continental plate drifted north, but the Arctic Platform and the Canadian Shield have been stable since the Precambrian; the gneiss and granite basement with its stack of sediments on top are still intact over hundreds of square miles, from Devon west over to Cornwallis Island, north to Ellesmere and south to Baffin Island. Even the various ice ages didn't substantially disturb the mud we're now trying to pick our way around. Usually when we think of ice ages, we picture glaciers moving aggressively down from the north and scraping the land bare to rock, as in Maine, or the ice thrusting out of the mountains, such as happened in the Sierra Nevada. Here, snow accumulated and the ice grew deeper, but it didn't move much. When the ice ages ended, the ice melted and outflow floods moved rocks around, but the glaciers didn't substantially rearrange the surface. As a result, there is this unusually long stretch of consistent geology, which makes it easy to compare with the inside of the postimpact crater created thirty-eight million years ago.

Impact craters are the most dominant landform on planets in the solar system. Not every planet has volcanism or tectonics, lacking sufficiently hot cores to create volcanoes or move crustal plates around, but every planet is bombarded by debris orbiting around the sun. We look at Mars and see volcanoes higher than Mount Everest, and canyons deeper than the Grand Canyon, but the most numerous large features are impact craters. When you gaze upon the severely scarred face of the moon, remember that the earth has suffered far more impacts than its satellite. The difference is that virtually all of our impact craters have been subsumed by tectonic recycling of the planet's crust during the last 4.3 billion years, or have been eroded away, or are now covered by oceans, lakes, and vegetation. The constant erasure of our planet's surface means that we've been able to identify only 160 or so such craters on Earth.

Haughton Crater is an anomalous feature in the High Arctic, a circular area of lighter ground that shows up clearly in satellite images. The largest impact structures on Earth are on the order of Chicxulub, buried in the Yucatan Peninsula, where a projectile six miles in diameter hit there sixty-five million years ago and blew out a ring structure 112 miles in diameter. The impact was so large that it wiped out 75 percent of all species on the planet, first by creating global tsunamis and wildfires, then shrouding the planet in a year-long night and lowering the global temperature ten to fifteen degrees. A relatively modest impact feature is Meteor Crater in Arizona, less than a mile in diameter and dating from only fifty thousand years ago. Haughton is a midrange hole, its twelve-mile-wide depression resulting from the impact of either an asteroid or comet up to a mile in diameter entering the earth's atmosphere at around thirty-three thousand miles per hour.

Haughton is the second highest latitude impact crater and the only known one exposed in a polar desert. Its circular form was first noticed in the 1950s by H. R. Grenier of the Geological Survey of Canada as he was examining aerial photographs of the islands. Grenier, naming the feature for a nineteenth-century British geologist who had studied specimens from Lancaster Sound, thought that it was probably a salt dome. It wasn't until 1972 that another Canadian geologist, M. R. Dence, suggested that it might be an impact crater. When scientists landed at the crater rim via helicopter two years later, they found confirmation on the ground that an impact had created the ancient devastation they saw before them. Shatter cones are gray rocks found only in fragments, and what they display no matter their size, from that of a thumbnail up to that of a football, are lines that radiate outward from multiple points. The rocks were shocked so severely that they didn't just melt, but were reordered at the molecular level by a process we still don't understand. They are found at only two kinds of sites—where nuclear explosions have been detonated and where the earth has been struck by a large object from outside the atmosphere. No natural earthly force is capable of producing shatter cones. The Canadians found them to be plentiful at their landing site.

When the asteroid or comet that created Haughton Crater hit the planet's surface, it was moving so fast and with such energy that it penetrated the surface by a mile or two and vaporized with the force of ten million Hiroshima bombs. The projectile generated temperatures of several thousand degrees and pressures equivalent to millions of atmospheres, sterilizing Devon in a flash of light. The shock wave depressed the surface of the planet more than six thousand feet deep for miles in every direction. That was the compression stage. Within seconds a refraction shock spread outward, releasing the tension, and blowing seventy to one hundred billion tons of dust, carbon dioxide, and sulfur into the atmosphere, creating a crater fifteen miles wide and lowering the temperature of the planet by two or three degrees for a year. By the time the Canadian scientists arrived, erosion had stabilized the crater walls into the twelve-and-a-half-mile-wide basin we're driving through.

The other evidence of impact that they found is shocked gneiss. Gneiss underlies the six thousand feet of marine sediments here, and is a basement rock of the North American continent, a dense, dark material that is harder and heavier than granite. It's closely related to the schist that sits at the bottom of the Grand Canyon and prevents the Colorado River from continuing to cut downward after excavating one thousand cubic miles of softer rocks during the last 5.3 million years. Gneiss is even denser and more obdurate . . . and yet, when Oz once put a chunk of gneiss into my hand that had been shocked during the impact, I almost dropped it in surprise. It was ash white, porous, and lighter than pumice. Many of its constituent minerals had been turned into gas by the impact and blown out of it. Ghost rock, I thought.

So there was a flash of light, then pressure and heat, an enormous dust cloud and a rain of rock. Imagine a nuclear explosion on serious steroids. Much of the crater is covered in silvery breccia, small rocks composed of the sediments that were blown apart and melted, then rewelded into rubble. The hole in the ground was so deep and steep that within minutes it began to collapse in upon itself. The blast had created an extensive series of fractures and faults deep within the surrounding rocks, and it opened hydrothermal vents that would soon bring water into the sterilized crater.

It took a few thousand years, but once the pool of molten rock cooled enough to allow the water to stand, the depression became a series of warm lakes that then, in turn, took tens of thousands of years to cool off even further. In the process, it became a new habitat for life. Spores blew in from the south, heat-loving thermophilic bacteria took root, and eventually fish returned. It was, in short, the kind of habitat that might have existed for millions of years on the surface of Mars when it was a much younger planet, and its core still had enough heat to produce volcanic activity.

When we get to the gully where Oz found the skull, within minutes Keith finds the embedded skeleton of a smeltlike fish that swam here after the impact. We work our way slowly upstream finding more fossils—the impressions of shells and twigs—until Oz locates an outcropping of rocks buried in the silts, out of which the fossils are slowly being washed away. We'd stopped several times along the way to investigate hillsides where the sediments were exposed, and dug down to see if any fossils appeared, but had hit permafrost within a foot and a half. Both the lake bed sediments and the Haughton breccia are permeated with ground ice, the breccia being a close analog to the Mars regolith, or surface rubble left over from impacts. And what we're doing here, roving the bed of a large crater to look for fossils, is the sort of EVA, or extravehicular activity, that astronauts will perform on Mars.

Permafrost in the High Arctic can go down more than two thousand feet before the crust warms up enough to melt it. The ice dates back one hundred thousand years to the beginning of an ice age that at its maximum coverage created an ice sheet as far south as the upper Mississippi River Valley eighteen thousand years ago. The ice sheet had retreated from Canada by six thousand years ago, although a large ice cap still covers the eastern third of Devon. Lift up the ground anywhere on the island and the ice age is right there at your fingertips, an ice so dark and hard that it takes a chain saw to cut into it.

But here the water has done the cutting for us, and exposed a stack of dolomite plates. Embedded in them are lenses of mineral that have accreted around the shells and bones of fish and anything else organic. Walk around camp, and you find ancient corals and the distinctive carapaces of trilobites from when Devon floated in the tropics. Preimpact fossils. Go down inside the crater and you find postimpact remains.

Oz walks upstream to make sure that more fossils aren't scattered about, then returns to the exposed rocks on one side of the stream and begins to pry them up briskly with his rock hammer. He excavates and sets aside enough samples to take back to camp. When he's finished he reminds me how Victorian collectors raided the Devonian sandstones in Scotland for nodules like these that they would then crack open for the fish inside.

While Oz heads downstream to confer with John about what he thinks might be some fossilized wood, I sit on one of the large rocks washed out of the glacial ice when it melted, and stare at an ejecta block across the Rhinoceros Creek valley. It's the size of a two-story townhouse and has tumbled down from the rim where it was thrown during the impact. I'm startled when a butterfly flutters by, the first I've ever seen on the island. The contrast between the apocalyptic force of the impact and the delicacy of the airborne insect stretches my comprehension of the Arctic in two directions at the same time, a familiar cognitive problem. Accepting the two is a matter of scale.

The diminutive Arctic willow is the tallest tree on Devon Island. Human beings evolved in the African savanna where we developed a sense of scale based on trees scattered throughout a rolling landscape of forest bordering open grasslands and waterholes. In the distance were often mountains, hazy and blue. We calculated how far it was from us to the nearest tree, and the potential safety its branches offered from predators, by instinctively and automatically measuring the length of our limbs with that of the foliage. We understood how far the mountains were, and thus could calculate how fast weather was moving in, by knowing how the spectrum shifted over the miles. Humidity scattered the light progressively the farther the mountains were, and we knew what each step along the color scale meant. We still do. And we still, most of us, prefer landscapes that look like that ancestral homeland.

Think of a typical urban park. Trees, grass, ponds. Think of your living room. A picture window with curtains framing the view. You can conceal yourself behind the frame and look out, safely unseen while you watch what's happening on a grassy lawn with trees; if we're lucky, we have a longer view across the street. The better that view, which is to say the longer it goes, the more expensive the real estate. Now consider the kind of art most people around the world prefer to hang on their walls: classical landscapes with dark foliage on both sides in the foreground that frames a view of the middle ground where there's a water feature—a stream or pond—and people or animals nearby. Mountains in the background, hazy and blue. We call this "scenery," which is also what we label the setting for plays in the theater (again, framed with curtains). We label what we like in the landscape "scenic." All are environments where we understand our place in things, place as in location and as within the hierarchy of nature, and even within our local societal group.

These parameters are, so to speak, out the window in deserts, hot or cold, the High Arctic among them. In the deserts we have the ground at our immediate feet and the far horizon as reference points, and precious little in the middle ground against which we can measure ourselves and the world. The air is so dry that it remains clear for miles, and the spectrum slides into the blue much more slowly than we expect. As a result, things look closer than we think they are. Even when we're aware of the effect, we tend to overestimate the distances to compensate. Trees the size of matchbooks, like the Arctic willow, don't help matters.

We can learn how to cope, either as individuals or as cultures; both Arctic explorers and Inuit, for example, have learned how to keep track of themselves in the very disorienting polar desert, and how to establish physical scale. It's part of how we change what is simply land, when we first experience it, into landscape, which is what space becomes when we convert it mentally into place. But that process here, which comes to us naturally in temperate environments, takes effort and patience, and makes Devon an excellent place to examine how our cognition interacts with the space around us. Which is going to be a real issue on Mars.

Imagine being on a planet where your very sense of the horizon is skewed because the diameter of the planet is only about half that of home, making the horizon that much closer. On Earth, you cannot directly see the curvature of the planet until you are eighty thousand feet high and almost out of the atmosphere. On Mars there are places, such as on its volcanoes, where you can see the curvature from the surface. Now imagine what it would be like looking through the darkened visor of a helmet while cut off from the environment. No wind on your skin to help you remember the direction in which you were walking. No sounds of running water to orient you to a watercourse, just the whir of the fan recirculating the air in your pressure suit. No touch. No smell.

The ability to establish our physical scale in the landscape is a cognitive skill that is so old and essential that it seems fair to speculate that it is hardwired into our perceptual systems. The ability to understand scale in time likewise has a genetic basis, but it is matched to the specific survival needs of individuals. Human beings can't perceive two events as separate unless they occur more than thirty milliseconds apart. (A common household camera can stop action down to a single millisecond, one one-thousandth of a second, at which point virtually all human motion appears to be frozen.) Even our ears, which have many times more nerve endings than our visual system, and which are exquisitely able to distinguish microtonal intervals in music, can't separate phonemes if they occur more quickly than that. Speech simply becomes unintelligible noise. This threshold is based on the frequency at which proteins in the brain can switch on and off—it's the lower biological limit to our perception of duration.

Within two seconds, if we are to continue to hold something in mind, we can no longer perceive a stimulus directly—part of the triage performed by the brain so it can continue to process new input—and we must invoke short-term memory. After three seconds, the brain starts actively seeking a change in the nature of the stimulus. Present a person with one of those well-known ambiguous drawings, such as the black-and-white outline of two profiles facing each other that is also the outline of a vase, and the mind flips back and forth between the two versions every three seconds, as if to query the world for news of change. Of difference.

We are able to hold many long-term personal memories in mind for the duration of our lives, but when we try to imagine the effect we have beyond more than the generation of our children, and perhaps our grandchildren, we are dependent upon survival instincts that function on the species level. For example, that template of landscape as park: We instinctively recognize that as a healthy habitat. When landscape looks ugly to us, we think something is wrong—it's a wilderness, or worse, a wasteland. We have a built-in aesthetic that helps us identify and preserve the natural environments with which we are comfortable (an aesthetic that, unfortunately, leads to inappropriate attempts to change deserts into well-watered grasslands).

When we need to consider the temporal scale of the geological process on Earth, most of us are lost once we are confronted with evidence of events more than a few generations old. Think how difficult it is going to be on Mars, where we will be confronted with a planetary desert, and where the surface has not changed much in billions of years. Its orange dunes and volcanic rocks will feel more ancient than we can comprehend, and its lesser gravity means that even ordinary movements will proceed at a different pace. It will feel unalterably strange.

In terms of the forces around us, we again scale everything to ourselves. We understand how much force it takes to lift up a glass of water or walk up stairs, yet most of us can barely imagine what it takes a weight lifter to pull four hundred pounds off the ground, or a runner to do a mile in less than four minutes. We're already far beyond our imagination when we grapple with the horsepower in our cars—not being able to remember how much weight your average draft animal can pull—although we know what it means to go twenty versus sixty miles per hour.

This is why the only way I can imagine the explosive force of the impact at Haughton is by holding what should be a very heavy rock in my hand and finding it improbably light, or seeing the force lines in a shatter cone put there by a means we can't identify. Pondering these matters while conducting an analog EVA is all the more useful because it foregrounds what we have to prepare for when we go to an alien planet.

I check my watch and it's five thirty, time to head back for dinner. We pack up the specimens, remount the ATVs, and manage to avoid most of the mud on the way back to camp. Once there we strip off our boots and parkas, and revel in the smell of the homemade corn chowder that Melanie Howell, our cook from Resolute, is serving tonight. Most of Devon Island is essentially odorless, there is so little organic matter on it. It's another small detail that reminds me of how many environmental clues we will be forced to forgo when we walk on the surface of Mars. John wouldn't be tasting Martian rocks or sniffing them for traces of hydrocarbons; he would be analyzing them with a spectrometer under sterile conditions meant to keep us from contaminating the environment—and to keep us safe, should there be organic traces to be found. Doing science without the use of our full panoply of senses will be a challenge far stickier than pulling ourselves out of the mud.

CHAPTER FOUR
Areology

Areo-, f. Gr. of Ares or Mars; *esp*. in astronomical terms relating to the planet Mars; as Areocentric *a.*, having Mars as centre. Areographer, one who describes the appearance of Mars. Areographic *a.*, pertaining to areography. Areography, description of the physical features of Mars. Areology, scientific investigation of the substance of Mars.

1877 D. GILL in *Mem. R. A. S.* XLVI. 94 The areocentric angle between the Earth and the Sun. 1878 NEWCOMB *Pop. Astron.* 566 Hourly motion in areocentric longitude. 1880 *Nature* XXI. 213 The local indistinctness and confusion that so often puzzle the areographer. 1870 PROCTOR *Other Worlds* ix. 93 The Martial geography—or perhaps I ought rather to say areography. 1881 — *Poetry Astron.* viii. 288 Compare. geology with areology

—The Oxford English Dictionary. 2nd ed. 1989. OED Online. Oxford University Press.

FOLLOWING DINNER and the inevitable hour-long bout of dishwashing for two-dozen place settings, one of the scientists or engineers usually gives a talk about what they're working on, plugging their laptops into a digital projector to illustrate comments with charts and photographs.

John has remarked often about how useful and reinvigorating it is to work with other scientists in the field here, whether it's another geologist such as Oz, or the team's microbiologist, Charlie Cockell. But science has gotten so complicated during the last century that it has become increasingly visual by necessity. When you're talking about the esoterics of impact craters or astrobiology to scientists outside your field, not to mention laypeople, it's useful to have a pictorial interface to summarize the data and bridge the gaps in understanding.

After the lecture—which tonight featured John talking about the relevance of water and hydrocarbons in rocks to astrobiology—a DVD is often popped into the laptop for a communal movie, which is usually when I slip away for a five-minute walk over to Fortress Rock. Last year John had walked over to it with me one evening and broken off a piece, then shoved it under my nose. It smelled faintly of rotten eggs, the telltale odor of hydrogen sulfide gas produced millions of years ago by bacteria decaying in an oxygen-depleted environment. He was grinning from ear to ear. But tonight I'm alone. The sun is low in the west and the temperature has dropped below freezing. It's an easy scramble to the top of the forty-foot-high crag of dolomite that stands east of our small dirt runway, but one made more interesting by heavy boots, bulky parka, and gloves. Should you break a leg, a medical evacuation would have to be organized to the clinic in Resolute. I take my time on the loose blocks and crumbling layers of rock, which at one spot are so thin you can peer through a crack to the other side.

At the top I do a slow 360 before sitting on the topmost slab. I've tried photographing a panorama from here, a pictorial technique first used on expeditions by painters in the eighteenth century and then later by photographers who wanted to capture and re-present an exotic scene. But I can't make one that satisfies me. Photos reduce all of the information present in a four-dimensional landscape to one flat surface that captures only an instant of time. My eye, by contrast, roves around from the foreground out to the horizon, left to right, top to bottom, inexorably seeking any changes within that period of every three seconds. A painted panorama gives us a slightly three-dimensional surface, and the artist will emphasize important details in the distance just enough to evoke the drama of a real landscape; a photograph presents a much more homogenized and less interesting, less resonant view. In a landscape as still as this one, the mind needs every kind of focusing technique available to it in order to sort out distance, scale, and direction. A painting would work better here than most photographs.

As the anthropologist Gregory Bateson put it: "All information is news of difference." He was mathematically correct both in terms of how we convey information and also in terms of functional neurophysiology. The brain scans the landscape for anything that moves, lest it prove a threat—a very old survival habit. As I rotate atop the crag, I can see nothing in motion—not a cloud, a stream, a person. Not a tree branch, a mouse, a flag. It is as still as an evening on Mars.

The human eye, descended from primitive cells 1.5 billion years old, is so sensitive that it can detect a single photon of light. The electrical activity inside the eye, however, generates random photons, so the brain has evolved a threshold of several flashes, five or so photons, before it registers light and calls our attention to it. This sensory triage is typical of the brain. Our eyes take in so much information, one hundred million bits every second, that the brain simply dumps most of it, lest it burn out. It does this by very quickly processing the input in comparison with patterns, some inherited, some learned. Everything we see in the world is constructed out of as few as two dozen simple geometric figures—circles, squares, triangles, and so forth.

The roving movements of my eye, "saccades" as they are called by neurophysiologists, allow me to notice the differences between brighter and darker areas of light in the landscape, to seek continuous lines, and to then compare the shapes thus created to what I know about the world. If something that is moving displays bilaterally symmetrical contours, I reflexively divert my attention to it. For example, an animal with two eyes placed on opposing sides of a nose implies the ability to coordinate vision and movement at speed, a hallmark both of predators and competing primates.

It's a brilliant piece of cognitive engineering that has made humans supremely gifted at pattern recognition, but it's also what led Lowell to mistakenly create canals where there were none. Nonetheless, the visual selectivity allows me to focus in on certain aspects of the world around me, to mentally magnify them. In a photograph, everything is small, on the same level of importance, and a relative hierarchy of values and information is obtained only with careful scrutiny. Looking at photographs is so foreign to how we see the world unmediated that in the mid-nineteenth century, when the technology was first introduced, people complained of "photographic distortion" between objects near and far, a cognitive dissonance that passed only when photographs became so widespread that everyone learned how to read them.

So I look around from my perch and try to build a memory instead of taking snapshots. To the north and just below is the dusty airstrip where the sturdy Twin Otter aircraft land and take off, ferrying scientists and supplies the hundred miles from Resolute. The runway is a flat area cleared of stones, not much larger than a football field and prone to vicious crosswinds. The ground slopes downward on the other side to the creek, then upward toward the plateau that Pascal has named the von Braun Planitia. Many of the local features bear names prominent in the history of Mars exploration, and Wernher von Braun, the German rocket scientist who helped build America's space program, was an early proponent of manned missions to the planet.

To the west from my vantage point is the HMP camp, its core formed by the long white kitchen tent where we eat, the two smaller orange tents for communications and lab work, and the nearly completed greenhouse built of bulletproof translucent plastic. The two dozen mountaineering tents in which we sleep are pegged down a few hundred yards farther west, well away from the food and any likely polar bear incursion.

Looking southward, a water line goes over the hill to a portable generator and pump by a stream (the Lowell Canal), and cables run to the top of a ridge where our satellite dish stands. The cylindrical white habitat of the Mars Society's Flashline Station is behind me on the northwest rim of Haughton Crater. Most of the rock on Devon is brown, but on the other side of the hab the landscape falls away into a silvery bowl of shocked rocks. When out on the ATVs during a field traverse in this landscape, it's easy enough to imagine you're driving on Mars. Fall behind the group a little ways until everyone else rounds a bend in one of the meltwater channels, and the sere landscape doesn't look like anywhere else on Earth.

It's never dark at this time of year, but from midnight until five or so in the morning, the light is low enough that your mind and body begin to slow. It's in those hours that the idea of driving on Mars tends to reverse itself into a question: What drives us to explore Mars? Carl Sagan, the astronomer who did much to popularize the search for extraterrestrial life, said that it was because its surface, with its polar icecaps and seasonal changes, seemed from a distance to mirror that of Earth. Pascal thinks it's also due to the fact that the Red Planet is actually orange, not red, a warm color we find welcoming. I think it's because we're seeking news of difference as well.

The first reason humans were attracted to Mars is its anomalous appearance to the naked eye. Humans in ancient cultures named and organized most of the brighter stars in the sky into constellations millennia ago (again, that irresistible urge to connect the dots). Five thousand years ago priests in Mesopotamia and Egypt, seeking to predict the future through celestial signs, noticed that five of the brighter objects in the night sky seemed to wander, even to reverse course from time to time. The wayward stars, which the Greeks named *planetes,* or wanderers, were Mercury, Venus, Mars, Jupiter, and Saturn, the five planets we can see with the naked eye. Only one of them has any noticeable color—the reddish tint of Mars. The Sumerians identified both the color and the planet with strife and carnage, an association also made by the Greeks, who named the planet Ares for their god of war. The Romans called the planet Mars for the father of Romulus and Remus, a diety who was also the patron of the Roman army.

The planet is unusual not only in its color but also in the way it brightens and dims on a regular basis. Egyptian and Greek astronomers, from Hipparchus in the second century BC to Ptolemy four hundred years later, proposed elaborate spherical celestial mechanics to explain its behavior. It wasn't until Johannes Kepler proposed in 1609 that the planet moved in an elliptical instead of a circular orbit that its retrograde motion in the sky was explained. That same year Galileo became the first person to look at Mars through a telescope, but his twenty-power instrument that peered at the planet when it was farthest in its orbit from Earth couldn't distinguish any features. By 1636, however, the Italian Francisco Fontana had a better instrument, and he made a sketch of Mars that showed variations in color. In 1659 Christian Huygens was able to observe with fifty-power magnification that Mars rotated on a north-south axis with a periodicity similar to that of Earth's rotation. Giovanni Cassini, working with Huygens, sketched Mars's polar ice caps in 1666. His nephew then discovered that the ice caps changed size on a regular basis.

The English astronomer Sir William Herschel discovered another important similarity to Earth, that the axis of Mars was inclined at roughly twenty-five degrees, only about a degree and a half more than ours—which meant that Mars had four seasons. This led Herschel to conclude in 1783 that the changing Martian ice caps were made of snow and ice, and that the dark areas on the planet were seas. The Germans Wilhelm Beer and Johann H. von Mädler compiled a map of the entire planet in 1840, which extended a network of supposed seas and arcing lines across the red globe, setting the stage for Schiaparelli and his *canali*. Furthermore, two other German scientists using a spectroscope in 1859 showed that the chemical composition of bodies in space contained the same elements as those found on Earth. That was the same year that Darwin's *Origin of the Species* was published and work was begun on the Suez Canal, both events seminal to the idea that there could be a civilization that had evolved on Mars.

Prior to the mid-1800s the moon had been the favorite destination of astronomers and early science romance writers, who proposed that alien races, perhaps of superior intelligence, lived there. Kepler himself drafted a book about traveling to the moon via magic, and as early as 1640 Bishop John Wilkins published his treatise *A Discourse Concerning a New Planet* in which he stated that the moon might be inhabited, and that we might travel there on what he called a "flying chariot" so that the British might establish a colony on the lunar surface. Once it became widely known that the small body was airless and frigid, however, Mars became the blank slate upon which to inscribe fantasies of interplanetary travel, authors spurred on by the discoveries and speculations of scientists. In 1864 the French astronomer and popular author Camille Flammarion stated that life must have evolved on other planets, including Mars, and in 1873 he proposed that its dark areas were, in fact, green vegetation. Following the flurry of European and American observations during 1877, made during the closest approach of Earth and Mars that century, which is when the supposed *canali* were observed, even major newspapers such as the *New York Times* began to wonder if intelligence weren't in residence on the Red Planet.

When Lowell declaimed the existence of a Martian civilization in his three books (despite the objections of Bernard, Hale, and Maunder) he was unconsciously expressing—and then magnifying, as part of a cultural feedback loop—the powerful public need for a new terra incognita, an unknown terrain that could be colonized by mind and body. For centuries that need had been fulfilled by Africa and the New World, and then, before they were mapped, the polar regions. In the early nineteenth century some explorers believed that the Antarctic might be a temperate or perhaps torrid land containing lost civilizations and unknown creatures. Even after Amundsen and Scott had reached the South Pole in 1911, Admiral Richard Byrd, preparing to make the first flight over it in the late 1920s, proclaimed in a magazine article that there could be life with prehistoric links still undiscovered on the continent.

Soon after the turn of the century, however, the mystery of the polar regions was beginning to lose its grip on the public imagination, as had earlier the moon, and Lowell's books helped Mars become the backdrop for legions of new science-fiction novels by authors such as H. G. Wells and Edgar Rice Burroughs. So widespread and persistent was the belief that life might exist on Mars that, when the planet again approached Earth in 1922 and 1924, the U.S. government asked radio stations to stop broadcasting, and ordered the cessation of all military radio traffic for three days in order to listen for communications from the planet. A master code-breaker was kept on standby to translate the signals. It is a deeply ingrained human trait to look for evidence of the "other," whether that means rival primates on the African savanna, foreign hominids such as the Neanderthals in ice-age Europe, or aliens in the solar system.

The first half of the twentieth century saw most scientific opinion, if not that of the public, come around to the view that Mars was yet another cold desert devoid of life, at least on the surface. Spectroscopic analysis of the planet starting in the 1920s showed the average surface temperature of Mars to be minus 40°F (versus plus 59°F on Earth), and that the atmosphere itself was impossibly thin. Starting in the 1930s, additional analysis showed that what had been called the "wave of darkening," which spread every spring from the poles toward the equator, was not vegetation springing to life with meltwater, but a change in the reflective properties of the Martian dust under seasonal conditions.

In 1957 the Soviets launched the 183-pound *Sputnik 1*, the first artificial satellite to orbit the planet. The event signaled the beginning of a race to conquer the high ground of space in the name of national interests, a space race which continues today, as exemplified by the alarm expressed by the American military when China announced plans in June 2003 to land on Mars (and return to Earth). Several months later China sent its first astronaut into Earth orbit. When Pascal and I were having our first meeting prior to my leaving for the Antarctic, he jokingly told me: "The only reason we'll go to Mars will be if China threatens to go there first." Early in 2004 President George W. Bush set forth the current mission for NASA to send manned missions first back to the moon and then on to Mars.

Much has been made of the race to put humans on the moon, but Mars has also been the focus of an intense competition, and one that remains centered around finding life—if no longer the traces of an ancient civilization, then at least microbes or their fossilized remains that will tell us that we're "not alone in the universe." The singular question in the minds of astrobiologists these days, if microbes are found on the planet, is whether or not their genetic material is related to that of life on Earth. No matter what the answer found on the Red Planet, it will have profound meaning to us.

The elliptical orbit of Mars sends the planet out as far as 149 million miles away from the sun, then brings it in as close as 124 million miles. Earth orbits in a nearly circular path at ninety-three million miles from the center of the solar system. When the earth, the sun, and Mars all line up, thus creating an opposition every 780 days, the distances between the two planets can be as little as thirty-three million miles or as much as sixty million. The former is called a favorable opposition, which comes around every sixteen years, and that is what occurred in 1877. As soon as humans had rockets powerful and accurate enough to reach Mars during an opposition, favorable or not, we started sending them up almost every two years.

The Soviets continued to outrace the Americans in space when in 1960 they launched two probes bound for the Red Planet. The Communist's Red Party wasn't about to pass up the opportunity to make a militant technological metaphor. Although both failed to attain even Earth orbit, President John F. Kennedy called for a manned mission to the moon in 1961, and mentioned sending people to Mars. The following year the Soviets launched three more vehicles toward Mars. All failed. That year we managed to get *Mariner 1* and *Mariner 2* headed to Venus, a closer and easier target than Mars, and although the first one plunked into the Atlantic, the second managed a flyby. Following the pattern of sending up two craft at a time every other year, *Mariner 3* and *Mariner 4* were launched for Mars next. The former failed shortly after takeoff, but *Mariner 4,* after a flight of seven and a half months, managed to fly within sixty-two hundred miles of Mars. During two days in mid-July 1965 it sent back twenty-two grainy black-and-white pictures that covered about 1 percent of the planet's surface.

Joseph Boyce, NASA's Mars exploration program scientist from 1985 to 2000, remembers the moment in the *Smithsonian Book of Mars:* "As the first photograph slowly filled the television monitors in Mission Control, scientists were both stunned and disappointed . . . where were the canals? Where was the water?" Although no one expected to find an ancient irrigation network, charts of the area still showed lines and sea basins. But instead, the dense layering of impact craters they were seeing looked less like the surface of the earth and more like the pictures of the moon that had been sent back by the Ranger missions the year before. *Mariner 4* used radio waves to measure the atmospheric pressure and found it to be less than 1 percent of Earth's—the gravity of Mars too weak to retain an atmosphere sufficiently dense for life as we know it. The spacecraft also failed to detect a magnetic field around Mars, which meant not only that its surface would be scoured by ultraviolet radiation every time the sun rose, but that it was being bombarded by high-energy solar particles that would destroy virtually any genetic material on the surface in short order.

The scientists were glum. The idea that Mars was a planet like Earth, or even one that could harbor life, seemed, for lack of a better word, dead, but the American and Soviet governments continued to fly spacecraft by Mars through the mid-1970s. In 1969 and only eleven days after Neil Armstrong walked on the moon, *Mariner 6* arrived to photograph the Martian equator, followed shortly by *Mariner 7,* which looked at the heavily cratered southern hemisphere. They passed within twenty-two hundred miles and sent back 202 photos that covered almost 10 percent of the planet's surface. NASA received more evidence of a chaotic aridity.

In 1971 both the Soviets and Americans again sent up dual spacecraft. The Soviet craft, *Mars 2* and *Mars 3,* were orbiter-lander combinations, meant to put vehicles on the planet, and both went into orbit. The *Mars 2* lander crashed, but the *Mars 3* lander reached the planet's surface and transmitted for twenty seconds before dying. It was the first signal to be sent back from the planet itself, and a remarkable achievement. *Mariner 8* failed to reach Earth orbit; but, as Oliver Morton put it in his 2002 book *Mapping Mars:* "*Mariner 9* became the first American spacecraft to go into orbit around another planet . . . the first mission to Mars to provide images of the entire surface and record the full diversity of the landscapes . . . to watch weather on another world."

The Soviet orbiters were on a fixed schedule of photographing the surface from high above the planet, most of which was unfortunately shrouded in an enormous dust storm when they arrived. The American craft, however, had been designed with enough flexibility that they could be shut down until the storm cleared, and for a month *Mariner 9* simply circled the planet. As the dust subsided, the cameras were turned back on. The first features to emerge out of the clouds were four dark circles, enormous mountains that appeared to have craters on their summits. Scientists back at the Jet Propulsion Laboratory (JPL) realized that these were not impact craters but volcanoes. The largest, Olympus Mons, rose sixteen miles above the average surface elevation of the planet (there being no existing sea level), and stretched more than four hundred miles across at its base. Its footprint was the size of Arizona, and at more than eighty-four thousand feet high, it was almost three times taller than Mount Everest. Although it was a shield volcano, thus more of a huge mound than a peak, it was the largest mountain in the solar system, a superlative geological feature that immediately caught attention in the press.

As if that weren't enough, the spacecraft took photographs of Valles Marineris, which turned out to be an equatorial canyon system that stretched nearly a fifth of the way around the planet. Perhaps "chasm" is a better word: It was an unimaginably huge gash 2,490 miles long, 420 miles wide in places, and up to six miles deep. Boyce points out that if it were on Earth, it would stretch from New York City to Los Angeles. The entire Grand Canyon—277 miles long, eighteen miles wide, and a mile deep—would fit into one of its tributaries.

The third and final major contribution made by *Mariner 9*, and one with direct bearing on the Haughton-Mars Project, was the discovery of the confluence of three valley networks into the low-lying basin of the Chryse Planitia. The channels, which were hundreds of miles long, debouched onto the plain and looked like nothing so much as the Channeled Scablands of Washington State. The Scablands are a maze of channels and islands sculpted when the Pleistocene Lake Missoula, which covered most of prehistoric western Montana, burst repeatedly through its ice dams. Over two and a half millennia those catastrophic collapses created the largest floods on Earth during the last two million years. At one point the floodwaters were two thousand feet high and carved their way clear to the Pacific Ocean. Chryse Planitia looked similar—except its channels would have been carved by outflows with a volume ten thousand times the flow of the Mississippi River.

The southern hemisphere of Mars was covered with ancient craters, but the northern hemisphere was a much younger, smoother highland shaped by volcanic activity and catastrophic flooding. Taken together, the three features meant that Mars had once been an active planet. Chryse Planitia might have been filled with a sea at one point, and would be a logical place to look for fossils. *Mariner 9* was only supposed to run for ninety days, but it kept transmitting data for almost a year, 7,329 pictures in all, about ten thousand times the total number of bits in visual information than gathered during the entire previous history of observation. Other instruments on board measured winds of up to 120 miles per hour, and determined that 90 percent of the atmosphere was carbon dioxide, but also contained clouds with water vapor. Before the mission, scientists had increasingly viewed Mars as a lunar-like environment. Now the pendulum was swinging back the other way.

American and Soviet scientists had been anxious to put landers on Mars before the images came back from *Mariner 9;* now they were even keener to conduct a search for life directly on the surface of the planet. NASA had hoped to put its two Viking craft on Mars in 1973, but the mission was delayed until 1975. In the meantime, the Soviets continued to throw spacecraft at both Venus and Mars. By the end of 1973 they had sent fourteen missions to Venus that failed, five that succeeded, and four to Mars, only one of which managed to orbit the planet and transmit pictures. Nineteen seventy-six, however, would bring a climax in Martian exploration that would last for the next twenty years.

The $1 billion Viking mission sent out two of the combination orbiter-landers that carried fourteen instruments each. The orbiters would photograph the planet and measure the water vapor in and temperature of its atmosphere. The landers would sample the atmosphere on the ground, analyze the chemical structure of the surface, and use six instruments to look for biology. In addition to two TV cameras to scan the terrain, they carried gas chromatograph–mass spectrometers to sniff out organic compounds, and three biology-sampling experiments to test soil. Based on Mariner photographs, and the subsequent maps of the planet made by the United States Geological Survey (USGS), one of the landers would be sent to Chryse Planitia, and the other would land at the edge of where the northern ice cap extended during its maximum seasonal phase. Both sites promised to hold evidence of water.

The Mariner pictures seemed to suggest that Mars had at one time been warmer and wetter than at present, and possibly a place where life could have evolved. The primary question about life on Mars was then, as now, whether or not the characteristics necessary for life to exist there were the same as or similar to those on Earth. Four years before the Viking launch, microbiologists had discovered bacteria living inside rocks from the Dry Valleys of the Antarctic, where minute quantities of water were suspended and the intense ultraviolet radiation of the polar regions was blocked. If life could exist on the harshest environment on Earth, perhaps it could survive the rigors of Mars.

The expedition took ten thousand people eight years to put together, but in September of 1975 the twin Viking rockets blasted off. Both landers made it to the surface of the northern hemisphere of Mars during the following summer and almost immediately began broadcasting pictures back to Earth. Carl Sagan—as well known for his public television series *Cosmos* as Flammarion had been in France for his popular astronomy books—was the director of the Laboratory for Planetary Studies at Cornell University, and he worked on both the Mariner and Viking missions. When he viewed the first pictures sent back from the surface by *Viking 1* at Chryse Planitia, he commented that it looked like the deserts of the American Southwest: sandy plains littered with volcanic rocks, interrupted by windswept sand dunes stretched as far as the camera could see.

At first the sky appeared blue until it was color corrected, and then it was a pale rose. Initial soil samples seemed to show a biological signal, but most scientists agree that they turned out to be oxidants that would be inimical to life. *Viking 2* found traces of methane and ammonia near the polar cap, but it was later determined that the results were caused by faulty interpretation. The soil contained a hundred times more sulfur than terrestrial dirt, and the atmosphere was 2 to 3 percent nitrogen, the rest carbon dioxide (Earth's is 80 percent nitrogen). The polar cap, which had been thought to consist of frozen carbon dioxide, turned out only to be coated with it, and instead consisted primarily of frozen water in a quantity that, if melted, would cover the entire planet to a depth of a centimeter. Furthermore, permafrost sat under the soil.

The two Viking landers functioned far beyond their ninety-day planned mission, eventually sending back forty-five hundred images, while the orbiters contributed fifty-two thousand. The mission results indicated that Mars was probably not now an environment suitable for life—but didn't prove it conclusively, as its tests were too limited to do so. And it still seemed as if the planet might have been warmer and wetter in the past, conditions under which life might have evolved. The unresolved questions could only be addressed by sending more landers, but Congress froze NASA's budget, and the agency was directed by President Nixon to shift its attention to developing the space shuttle.

The exploration of the Red Planet was on hold until the Soviets sent *Phobos 1* and *2* in 1988. The first of the orbiter-lander craft went out of communication with Earth while on its way; the second achieved orbit around Mars, but then it, too, lost touch with ground control.

Not to be outdone, in 1989 President George Bush, on the twentieth anniversary of the moon landing, called for a manned mission to Mars. NASA still had one launch in the pipeline, the billion-dollar *Mars Observer,* in essence a geoscience satellite converted to survey Mars, and it was sent up in 1992 to pave the way, only to disappear inexplicably. The first NASA loss of a complete mission—because only one craft was sent instead of two—led to some serious head-scratching.

Among other things, *Observer* was supposed to have provided high-resolution pictures of where the next lander could put down.

That year, while the NASA legend about a great galactic ghoul who lives between Earth and Mars and eats spacecraft once again made the rounds, a former businessman was appointed director of the agency. Dan Goldin decided that NASA would send twelve to twenty relatively inexpensive landers to the planet. The idea was that NASA would no longer send spacecraft in pairs, as had been done with Mariner and Viking, nor would redundancy be built into each system for every kind of imaginable failure. Instead, by relying on the increasing power of computers, the craft would have the flexibility to be programmed while in flight to recover from problems. It was what he called the "faster, better, cheaper" business model, and its product was Pathfinder, which combined three elements. The mission would test the concept of sending a landing craft directly to the surface of the planet without using the cumbersome orbiter-lander system; it would conduct meteorological observations; and, as a late addition to the mission design, it would deploy a small rover to attempt some geology.

The story of how the Pathfinder mission came together has had entire books written about it, but suffice it to say that NASA started with the fact that no one who worked there anymore had the faintest idea how to put a spacecraft on Mars. People were hired out of retirement, a test rover that had been toyed with was shrunk from fifty to fifteen pounds, and the Russian method of using a giant cluster of airbags to cushion the landing was adopted. In real dollars, the mission cost only a tenth of the Viking's price tag.

Pathfinder bounced up to fifteen times when it hit the dusty surface of Chryse Planitia on July 4, 1997, but the airbags worked, and shortly thereafter *Sojourner* became the first wheeled vehicle to go a-roving on Mars. The vehicle was a petite two-foot by one-and-a-half-foot box on six wheels covered with solar panels on top. It carried an alpha-proton x-ray spectrometer to identify elements in the Martian surface, and two crude cameras. Among other tasks, *Sojourner*'s job was to look at other Martian rocks for comparison with the meteorite from Mars that Robbie Score had found several years earlier in the Antarctic (and which had shown possible signs of fossilized nanobacteria in it). The outflow plains of the channels were yet again a logical place to look, as well as being a relatively level place to land.

Within days *Pathfinder* was sending back stereo color panoramic shots of a volcanic plain covered in rocks that had been transported from impact craters by a massive flood. The little rover circumnavigated the lander, looked at rocks on the orange plain, and found conglomerates that had been formed by water. That meant that at one point water was stable on the surface for millennia—not just in the atmosphere as water vapor or ice, and not just locked up underground as permafrost. Life could have arisen on Mars at roughly the same time as on Earth, more than three billion years ago, which increased the possibility that a Martian rock blasted off the planet by an impact could have, indeed, contained microbial life.

NASA posted the pictures they were receiving on the Internet: the barren Martian desert, dust devils, the pale gray sky of a Martian sunset. During the first month its Web site devoted to the mission logged more than 566 million hits, making it the largest Internet event to date. The rover had been designed with just enough onboard intelligence that it could avoid small obstacles and survive the twenty- to thirty-minute time lags in radio communication with JPL (Jet Propulsion Laboratory), thus proving you could drive on Mars with remote control. *Sojourner* lasted more than three months, but as winter approached it could no longer collect enough sunlight to maintain power, and it was shut down when the lander batteries finally failed. The mission ended in late 1997, having sent back more than ten thousand pictures.

In September of that year another American craft went into orbit around the planet. The *Mars Global Surveyor,* built primarily out of spare parts meant for the lost *Observer* craft, arrived and began to brake into its circular orbit for mapping the planet. It did so by dipping into the thin Martian atmosphere, a procedure that brought it close enough to the surface to detect the remnants of a magnetic field. The discovery meant that the planet could have been protected enough at one time from solar radiation that genetic material could have survived. Within its first year the spacecraft produced a better planetary topographical model than that available for Earth; it identified hematite deposits, and had begun to estimate water inventories on the planet. Its high-resolution cameras have since taken more than 150,000 images of the surface, many of which strongly suggest that numerous features were caused by flowing water. Now the issue is how much water remains frozen on the planet underground.

NASA followed up with four more spacecraft in 1998. The *Mars Climate Orbiter* failed to achieve the correct orbit around the planet and disappeared, one of its teams working in metric units, the other in English measurements. The innovative *Mars Polar Lander,* which carried two small probes in addition to its lander, apparently failed to fire its rockets for landing. Both were victims of the cheaper way of doing business, which was subsequently and substantially modified, and the *Mars Odyssey* was successfully put into Mars orbit in 2001.

By this time the Russians, who had seen their last Mars mission fail to achieve Earth orbit in 1996, were struggling just to fund their share of construction on the International Space Station. The Japanese launched their first mission to Mars in 1998, but it failed to achieve Mars orbit. Currently, only the European Space Agency has successfully accompanied the Americans to Mars, putting into orbit their *Mars Express* in the same season that *Spirit* and *Opportunity* landed on the surface. All the vehicles are working together to link communications, and with *Surveyor* and *Odyssey* are imaging the planet at a level never before achieved.

I've been sitting on Fortress Rock for almost an hour and am getting cold and stiff. It's not that the temperature has dropped by any significant degree, it's just the slow attrition of low light and cool temperatures to which I'm barely paying enough attention, and the slightest of breezes. If you sit absolutely still, a thin layer of warm air envelops your body and you can stay surprisingly warm for a long period of time. But even a slight movement of yourself or the air breaches the barrier. It's nothing you can see; it can only be felt.

Scientists used to talk about "picturing" nature, then "mapping" it. Now they tend to speak of "imaging," whether they are talking about the pictures made with scanning-tunneling microscopes of atoms or the depictions of large-scale structures in the universe at the farthest limits of the electromagnetic spectrum. It's a word that in terms of Mars includes hand-drawn sketches based on observations through a telescope, photographs from an orbiter, television images from a lander, and laser altimeter data sets—and making maps by all of the above means. But it also includes creative acts such as painting an imagined Martian landscape, writing novels about rescuing princesses from alien warriors, and composing film scores for science fiction flicks. We translate all of those into mental images.

Mars has been imaged more than any other surface off the planet, and in fact was mapped by *Mariner 9* and the United States Geological

Survey twenty-five years before it was possible to do so for the Antarctic, some part of which is always shrouded in the cyclonic storms constantly circling it. Today you can use the NASA data sets to create a three-dimensional model of the surface and fly virtually through the Valles Marineris, or around the great summit of Olympus Mons. But the data is still not the territory.

The photographs I've taken from atop Fortress Rock fail to satisfy me because they are not only visually reductive but also bereft of other sensory input. We place our bodies in space by haptic means, by using a full-body sensorium that uses smell, taste, hearing, and touch to establish our literal balance and orientation in a landscape. These are the very clues that Paul was teaching me while we snowmobiled across Cornwallis Island four months earlier. Part of my memories of being on Devon are the ultra-bright aridity of the air in my nose, the blue mineral taste of Arctic water, the boxing of my ears by wind on the tent. It's not simply an accumulation of multisensory perceptions, but a more complicated synesthesia, a complex overlayering and transposing of senses that are processed in and compared with memories and expressed via literary terms and other cultural analogs, such as metaphors (the root of which is Greek for bearing from one place to another, of transferring attributes from one object analogous to another).

To fully transform a space into a place, or a terrain into a territory—to experience land as a landscape—you have to be there in person and involve all of your senses, and then translate that experience to others. The land has to be able to hold meaning in our imagination, to evoke an emotional resonance, and it takes more than aerial images transmitted by a rover, although those are a start. It takes people on site collecting impressions, and then transforming them into stories and images that re-create for us how they felt when there.

To know Mars, we will have to feel the temperature differential between our feet and our heads, which on Mars can run more than forty degrees. We will have to feel the wind pushing us around—and on Mars, because the atmosphere is so thin, the 120-mile-per-hour gusts may only feel like ten miles per hour—yet, because the gravity is only a third that of Earth's, we'll be enveloped in a dust storm, which will be a profoundly disorienting experience. We will have to hear how our footsteps fade in air that is barely there, and listen as the powerful static of a dust devil brushes across our pressure suits.

We can image and imagine Mars all we want, but if it's important for us to know it, we have to go there in person. And then, in addition to cataloging and analyzing its rocks, we'll have to make paintings, poems, music, dance, sculpture, plays, and novels of it all, because these are the ways in which we translate sensory results from one person to another. If we want Mars to be more than space or alien terrain, if we want it to become a place within grasp of human society, we'll have to learn how to re-create our sensory impressions of it in art.

CHAPTER FIVE
Living Under a Rock

THE HAUGHTON-MARS Project has two ongoing interrelated programs, both of which are international and multidisciplinary by design. As Pascal puts it: "The exploration research allows people to study key technologies, strategies, hardware design, and human factors relevant to the future exploration of Mars and of other planets by robots and humans." Most of the daily activities on Devon have to do with exploration. But what drives that program is the science research, which follows three themes: What can we learn about Mars by comparing it to analog environments on Earth; what are the effects of impacts on Earth; and, how does life exist in extreme terrestrial environments? What pushes the science research is the question of life on Mars, and the HMP lead biologist addressing that question is Charlie Cockell, an affable thirty-six-year-old English microbiologist.

This morning Charlie and I have taken Ivan Semeniuk, a producer from Canada's Discovery Channel, and his videographer Hernan Morris out to the Breccia Hills so they can do an on-site interview with Charlie explaining how life came to colonize the crater, and what that has to do with finding life on Mars. Ivan is a quite sensible director and interviewer, and Charlie, who is in his midthirties and possessed of a dry wit that he intersperses without apparent effort, a good subject. I always enjoy wandering around the small ridge while Charlie pokes about on his hands and knees for samples of endolithic bacteria. You find the endoliths by turning over the rocks to see if they display a telltale green stain underneath, where the unicellular organisms are protected from the harsh weather and radiation of the Arctic, or by breaking open the rocks to see if they're in the cracks. I prefer to wander the silvery breccia looking for pieces of shatter cones, which never fail to fascinate me. Ivan, in an aside to me, says that he considers their radiating lines "the signature of sound in rock." I am very taken with the description, at once both literal and a metaphor.

Small yellow poppies and purple saxifrage flowers are also scattered widely across the breccia, along with green map and red bloodspot lichens. There's a bit more vegetation inside the crater than outside the rim, which is due in part to the interior being slightly sheltered by the topography, as well as benefiting from meltwater flowing inward. It's also because the shocked rocks provide a better postglacial habitat. The breccia itself is pretty much like cement fused out of rubble by the blast of the impact, but the shocked rocks like the gneiss have those empty interstices that are available for colonization, which could also be the case in and around the impact craters of Mars.

"I was here with a Japanese television crew filming extreme environments," Charlie tells us, "and when they saw the flowers, they got all upset because it didn't look extreme enough—so they started digging up all the plants!" Clearly they had been determined to present Devon Island to the media public as more of a literal Mars on Earth than an analog. Ivan shakes his head and Dr. Cockell shrugs his shoulders, then goes to work for the camera.

The first time I saw Charlie in the field was last year when Oz and Darlene Lim, a Canadian limnologist studying the small lakes and ponds in the crater, took me out on a short ATV traverse after dinner one evening to where he was working. Two of the teenage Inuit camp helpers from Resolute, Jeffrey Kheraj and Sandy Salluviniq, also mounted up on ATVs, and Darlene grabbed one of the shotguns hanging inside the kitchen tent. The first thing you do upon arriving in camp is put up your tent. The second, if you are a newcomer, is to receive polar bear training from John Schutt, who, as part of his responsibilities managing the camp, makes sure you know how to use the shotguns placed within strategic reach. After trudging up the hill behind camp, John has you practice shooting solid slugs from the twelve-gauge into a cardboard box several yards away.

The rules regarding the apex predator of the Arctic are as follows. Never leave camp without a radio and a shotgun, even for a short stroll. Most polar bears avoid the interior of Devon Island as there's precious little game on it for them to eat. The number of humans in July, perhaps sixty at most, about equals the declining population of caribou on the island, but . . . the coastline of the island is an important maternity and denning habitat for the great white bears. The occasional juvenile bear does traverse the island out of inexperience, and they do visit the camp from time to time, "poking a fingernail into a corner of the cook tent to see what's there," according to John. He patches the holes with duct tape. Said inexperience means they may not be very wary of humans, though they tend to shy away from the dog we keep in camp for that purpose.

More rules. If a polar bear paces back and forth in front of you, and then displays its left shoulder to you (they're left-handed), it's probably going to attack. If it comes close enough to you to hit it, shoot. Keep shooting until it falls down. If it's still coming at you when you run out of solid slugs (which, unless you're a trained hunter, is quite possible), there's a shell of birdshot left—maybe you can blind it. If perchance you are lucky enough to survive and the bear isn't, pull out your checkbook. You now owe the local Inuits $25,000 for the hunting license that they could have sold to a paying customer. This is, unfortunately, not an expense covered by NASA.

So, although no one here has yet to shoot more than warning shots at a bear, Darlene wrapped the shotgun in a plastic garbage bag and secured it across the front of the ATV with bungee cords, and off the five of us went at a sedate pace along the airstrip past Fortress Rock. We wound along and over the dusty rim past the Mars Society's habitat, then picked up a little speed as we descended the gentle slope into the crater. The crater bottom is a maze of small ridges and valleys, but our route followed the familiar track toward Rhinoceros Creek—straight across the depression, rightward into a petite valley and past the Breccia Hills, over a small saddle past a large ejecta block. Instead of heading right and down into the Rhinoceros Creek drainage, we bore left to a small rivulet where Charlie had set up camp.

Next to his red four-wheel ATV sat a rucksack with gear for collecting samples, his shotgun, and a tarp to wrap up in when it rained, which it was threatening to do. A series of moss samples sat in rows on the ground where they had been absorbing differing amounts of ultraviolet radiation, awaiting transport back to England for laboratory tests to assess the damage to their DNA. That was the extent of his camp. But, then, Charlie has long years of experience in making do with little in the field. When he was ten he took to launching model rockets from the lawn at his boarding school with houseflies as a payload. The water-pressure-propelled craft would go straight up two hundred feet in the air and then fall back earthward under a parachute. The flies lived. In 1990 when he was twenty-two he organized an expedition to Mongolia—the first western expedition to enter the country after the communist era—and crossed fifteen hundred miles of sand and mountains in order to study the long-eared hedgehogs of the Gobi Desert. Three years later he was leading a foursome into the rainforests of Sumatra in Indonesia in order to fly an ultralight at night just above the forest canopy to collect moths. Although he crashed the aircraft, totaling it but not himself, he managed to net ten thousand moths for study.

As a microbiologist with a PhD from Oxford in molecular biophysics, Charlie earned a staff position with the British Antarctic Survey in Cambridge, and has visited the Antarctic three times, as well as coming to Devon Island since 1998. He's also run for prime minister as the only candidate in the Forward to Mars party, and in 1996 managed to collect ninety-one votes. As has been the case with many field scientists throughout history, he's an amateur painter—specializing in notable scenes of the exploration of Mars, a history yet to occur. His most recent books include one on the ubiquity of microbial life through the four-billion-year history of life on Earth; a perfectly serious handbook for Mars explorers that covers such topics as what to put in your medical kit while driving on the Red Planet; and a third that demonstrates how terrestrial environmentalists and space explorers are increasingly and inextricably bound to one another through philosophy and science. All three books have benefited from his time on Devon—and to think that the handbook or the exploration paintings might be premature is to miss the point.

Charlie first met Pascal at Ames in 1997, the year that Pascal received his small grant from NASA to fly north to Haughton and check it out. Charlie's enthusiasm for exploration and astrobiology was a perfect fit with the work that needed to be done in the crater, and he was on the crew the next year. Charlie studies several aspects of terrestrial biology that are related to life on other planets—a field known as exobiology or, more widely these days, astrobiology—studies that include extremophiles and photobiology. Extremophiles can live under exceptional conditions such as polar cold, the heat of volcanic vents, intense radiation, and pressures ranging from bone-crushing at the bottom of the ocean to those on Mars, so low that they would leave you gasping for air. On Devon he spends most of his time looking at what lives in and under rocks, figuring out how life uses light for energy while surviving ultraviolet radiation and other hazards of the electromagnetic spectrum.

While Oz and John are out prowling the crater for samples and fossils, puzzling out the geological and biological histories of the impact structure, Charlie is teasing out the life that lives here now. Given that impact craters are the most ubiquitous features on solid surfaces in the solar system, if we can understand how life-forms colonize them, we will learn how to look for them on places such as Mars. The form of life that is the oldest and most stubborn on Earth is the microbe, a unicellular organism that can be as small as a fifth the size of the *E. coli* that live in the human digestive systems—which are only a thousandth of a millimeter in diameter, or even smaller. No one yet knows. They can also be as large as the forams that live in the ocean, the shells of which can measure up to a millimeter (about four hundredths of an inch). Colonies of dense microbial mats, as Charlie explains in his book *Impossible Extinction,* existed on Earth as early as three and a half billion years ago, and are found today at the bottom of the permanently frozen lakes of the Dry Valleys in the Antarctic, as well as in more hospitable climes. When these early microbes were growing, the atmosphere of the Earth had two thousand times the amount of carbon dioxide in it as now. Given the lack of oxygen and, therefore, an ozone layer, ultraviolet radiation would have been perhaps a thousand times stronger, as well. What the missions to Mars have shown us is that conditions on that planet may have been similar during the same time period.

Life on Earth was pretty much microbes all the way down for almost three billion years. Animals have existed only for the last six hundred million years or so, a late by-product of the oxygen that microbes started creating once they evolved the ability to photosynthesize energy from sunlight, which occurred perhaps two billion years ago. Animals evolved as organisms that require oxygen to live, not just to breathe, but actually in order to eat. They are ultimately dependent on other photosynthetic forms of life in order to sustain themselves (think of a food chain that starts with plants and ends up as everything from granola to hamburger in your stomach). Some microbes, however, can live off chemical reactions other than photosynthesis. They can live underwater and underground while they metabolize iron, for example, the fourth most common element on the planet. Microbes can eat rocks, each other, and just about anything else, and do so without using sunlight themselves. Scientists have sent them up into Earth orbit aboard a rocket for six years, where they were subjected to temperatures close to absolute zero (–459.67°F), zero gravity, total vacuum, and zero protection from radiation. Brought back to Earth the *Bacillus subtilis* were still viable—they bred. Microbes can survive the jolt of being blasted off one planet by a meteor and, if inside a rock, the heat of entering the atmosphere of another. In short, they can travel from planet to planet.

Eighty percent of the biomass on Earth today is comprised of microbes. As Charlie pointed out in one of his after-dinner lectures in the kitchen tent, you have one hundred thousand million microbes wandering around the one hundred billion cells of your body. Drill down fifteen hundred feet into the earth, and you'll find up to ten million of them in a space equivalent to a pencil eraser. Two miles down you'll find them eating hydrogen and carbon dioxide to survive. Forty miles up in the atmosphere? Microbes. At the bottom of the Atlantic under the pressure of 250 atmospheres and with temperatures well above those where water would boil at the surface? Microbes are metabolizing sulfuric minerals. As Charlie cataloged their adaptations, I remembered my surprise at finding microbial life everywhere in the Antarctic, even in Lake Vostok underneath three miles of ice, where they wouldn't have been exposed to the air for a million years.

Microbes are, in Charlie's eyes, the ideal candidates to survive the kinds of catastrophic impacts and volcanic eruptions that can kill off everything else on a planet. The primary reason that Charlie is on Devon is to study how microbes are doing in the crater, and the answer is, quite well. The shocked gneiss, for example—that black, dense, heavy material that has been turned into a relatively porous rock by the impact—is a fine habitat for microbes. By fracturing and vaporizing out minerals in the gneiss, the impact opened up space for them. Furthermore, it turned the rock translucent so that the microbes are protected from some of the ultraviolet radiation while still receiving sunlight for photosynthesis. Charlie, along with most of the other astrobiologists with whom I've spoken, agrees that evidence of microbes living in rocks—the photosynthetic cyanobacteria that are also found in the cold desert of the Dry Valleys, as well as in hot deserts such as the Negev in Israel—is what we should be looking for on Mars.

The increasing range of microbial habitats on Earth has raised the optimism of astrobiologists about the likelihood of life on Mars. Liquid water once existed on the planet, and was perhaps abundant and stable enough in large bodies to be favorable for life to have evolved from organic chemicals. So the first question to be addressed is, has life ever existed there? If yes, the second question is whether or not any of that life has survived until the present day. Third, if there's been life on Mars, did it share a genome with Earth? If genetic material has been shared between the two planets and there was a common genesis for life, that means we are in a sense Martians already. If there were two separate incidents of genesis, that raises the chances that life is more widespread in the universe as a whole. The answers to all three questions most likely lie in the absence or presence of microbial fossils on Mars.

While we were examining Charlie's samples of moss, Darlene motioned me over to a micro-oasis. Although 97 percent of Devon Island's ground is bare of vegetation—the ground was actually made sterile more from the effects of massive glaciation than the impact—when an animal dies its body provides nitrogen to fertilize the soil. This creates a small area in which seeds can germinate and plants grow, which in turn attracts small mammals, such as the lemmings that frequent the terrain. They burrow through the ground and their droppings further enrich the soil in a feedback cycle. Darlene had found a small seep of water with just such a patch, and she picked a piece of mountain sorrell for me to sample. Its intensely clean green taste reminded me of watercress. The poppies and saxifrage were growing in the oasis as well, and she got down on her hands and knees to photograph the miniature garden, then pick some of the sorrel to add to salads at camp.

When she was done we trooped up the hillside that separated us from the main valley of the crater, and as we crested the ridge the landscape widened and deepened all at once, a silvery bowl across which shadows chased. This was all Inuit Owned Land (IOL), and we were not allowed inside its boundaries without special permission from the governing council in Grise Fjord, a tiny settlement on Ellesmere Island even farther north than Devon. Pascal works every year to reassure the members of the council that the HMP personnel are focused exclusively on science, and are not looking for oil or diamonds or anything else of monetary value in the crater. Although everyone is sympathetic to preserving the lands for Inuit needs, the travel restrictions have sometimes been a source of anxiety for the scientists who need access to the crater, especially Oz, who couldn't complete his studies without their permission to work down inside of it.

Rhinoceros Creek ran below us, and bent out of sight to the northeast where it joined the Haughton River deep in the heart of the crater. In front of us were the hills pushed up by the rebound of the impact, a feature often seen in larger craters on the moon and Mars. I walked up to the highest point of our summit, a modest tabletop of fractured rock, and just beneath it found another micro-oasis. When Darlene joined me she poked about under its grassy mat and found the small bones of a bird that were still decomposing, and would be for years. The microbes work slowly here during the cold, short seasons.

Below us a long-tailed jaeger took off from a rock and flew down the creek, then banked across to a landing spot on the other side. It was the first bird I'd seen on the island, and I wasn't surprised that it was in the crater, versus outside it. We could just glimpse a small patch of tundra in the middle distance, biologically the richest terrain on the island. The series of steaming lakes created by the impact eventually drained through the rim of the crater, leaving behind their sediments. Most of the hardened mud eroded away, but there's still a four-mile square patch left in one of the lower parts of the crater. The sediments are relatively rich in nutrients, and its well-watered meadow is enough to support a family of Arctic foxes that Charlie had been watching, as well as some Arctic owls that were hunting the endemic lemmings, and a rarely glimpsed visiting herd of a half-dozen woolly musk oxen.

I gazed for long quiet moments at the creek, the far side of the crater with its cliffs, underneath which run the river, and the startlingly green acreage of Arctic grass a couple of miles to the southeast, which Pascal had named the Lowell Oasis. There was a small chance that Oz might be allowed to traverse the interior reaches of the crater this year, if accompanied by the obligatory Inuk guide, and that I could accompany them. The view made me mentally cross my fingers in hope.

Darlene and I were joined by Jeffrey, one of the teenagers from Resolute. A tall, slender, and somewhat dashing kid, his sole ambition in life seemed to consist of becoming nothing less than a Jedi knight. Pascal hires several Inuit high school students every July to work in the camp and learn what the scientists are doing. It's part of his effort to allay concerns of the council in Grise Fjord, and it also satisfies his own moral precepts, which dictate that you involve local people economically in whatever it is you're doing no matter where you are.

Jeffrey mounted the rocky summit and twirled about in imitation of a martial arts move, a dark-haired version of Luke Skywalker from *Star Wars*. Part of me was dismayed by how the younger Inuit listen to rap music and watch the latest Hollywood movies on DVD—a cultural exchange that we foster by our presence. That's not a new or unusual feeling for many Americans who travel to remote places these days. But part of me also grinned in bittersweet acknowledgment of how perfect it was that Jeffrey would adopt the legend of a young boy, destined for big things, growing up in a rural desert world. There we were, techno-wizards of an imperial power acting out our inscrutable roles on the analog for an alien planet, and he was a participant.

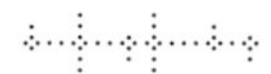

We all have our fictional tropes on Devon Island, and that evening last summer was the first time I began to realize how deeply embedded in them I was. The media representatives that visit both the HMP and Mars Society hab every summer only reinforce the notion. We've walked around and talked with reporters, writers, and photographers from American and Canadian outlets such as CNN, MSNBC, *National Geographic,* the *Los Angeles Times,* and even Russian National Television—and we've all willingly assumed the roles we've been assigned.

This morning, after Ivan and Hernan are finished with Charlie atop the Breccia Hills, we ride over to the ejecta block so they can film him taking samples. The various blocks around the crater serve several purposes for HMP members. First, they are landmarks. We use this one to recognize the key saddle on the route to camp when returning from the Rhinoceros Creek area. Second, as evidenced by my musings when out with Oz and Keith, they are almost sculptural reminders of geological importance and of the titanic forces that shaped this landscape. Third, they have become habitats for all sorts of life. Not only have endolithic bacteria settled into the cracks of the conglomerates and lichens colonized their surfaces, but sparrow-sized snow buntings have taken to pecking away at natural gas pockets in the rocks, enlarging them into foot-deep holes they live in, a most unexpected sight.

No birds are currently in residence at this block, and Charlie taps merrily away at its face with a rock hammer, popping out pieces as large as his thumb. Anything interesting goes into plastic sample bags, which are then sealed. "The endoliths only grow down to a centimeter or two; that's when they run out of light," he notes for the camera. He turns and gives Ivan a minilecture, gesturing with his hands, appearing to be very much the serious researcher in a remote corner of the world. And he is. I stay carefully out of the frame of the picture, keeping to my role of the writer observing from the outside—but I am within another frame of my own making, that of the narrative I am constructing out of three trips to Devon and the story of a rover on Mars.

CHAPTER SIX
Across von Braun Planitia

Joe Amarualik lifts a twenty-pound slab of dolomite and sets it sideways across two other pieces that rest on the ground. He steps back, tilts his head, leans forward to adjust the dark brown rock an inch to the left, then turns to the next rock we've brought him. He uses three dozen pieces in all before he has the legs done, each one just over three feet high and almost as large around. The sleet blowing sideways by us this morning slows to a drizzle, and we unzip our parkas as we scour the rocky knoll for more slabs.

Five of us have driven out across the arid von Braun Planitia to the farthest ridge visible from camp on what is the anniversary of both the *Apollo 11* landing on the moon, the mission that saw the first moon walk, and the arrival of *Viking 1* on Mars, the first successful landing on the Red Planet. We're erecting an *Inuksuk* on a small knoll above the saddle through which the Humvee drove the other day and from which its crew obtained the first view of camp. *Inuksuit* (plural for *Inuksuk*) are human-shaped cairns that can be as simple as one rock leaned against another, or, like the one next to me, can have two legs, a torso, arms, and a head—all sketched with an economy of gesture that a minimalist sculptor would envy.

Inuksuk stems from *Inuk,* the Inuit word for "person," and the figures literally stand in for human beings on the land. They can indicate a place of safe passage, commemorate a significant event, or be constructed just to keep a lonely hunter company. They often serve as landmarks in those wide empty places where the High Arctic offers the traveler few clues. On what we take to be an auspicious day, we're erecting this particular assemblage for Michael Anderson. It is the third of seven such cairns we hope to build this summer to memorialize each of the crew members of the space shuttle *Columbia.*

Pascal lugs over two more slabs, large ones that will be the prominent pelvic bones. No one knows how long people have been erecting the stone figures in the High Arctic, but a succession of cultures has moved through here for more than four millennia, and *Inuksuit* are as direct a transformation of land into landscape as can be imagined. Thousands of them are scattered throughout Nunavut, and they have become the official symbol of the territory. They are, however, rare on the uninhabited Devon Island, and Joe thinks that this will be the tallest one built here so far.

Most of Devon is an unfriendly maze of glacial trough valleys and dendritic meltwater channels in which it is easy to become lost. The navigational difficulties are magnified by the circling of a sun that never sets at this time of year—assuming you can see the sun, which for the last few days has been almost impossible. Even in warm years, such as this one, sea ice holds fast all year long to much of Devon's shoreline, and deep pockets of snow stand in the labyrinth of rocks behind camp. The snowbanks here are long-term residents, and when traveling across the island, we orient ourselves to prominent ones from the fifty-year-old aerial photographs that we carry.

Joe carefully positions the two pelvic slabs that Pascal has contributed, then four of us manhandle upward a massive chunk of the sharply pitted umber rock to begin the torso. After another half hour we find a triangular rock to sit atop as a head, making the figure more than six feet tall. Nature doesn't have a mechanism for stacking stones, and when you come across three or more sitting atop one another on flat ground, you know that human beings put them there. You can't see an *Inuksuk* on Devon and not detour to visit it, so powerful is our need to connect with human presence in a space this daunting.

Keith, who had gotten official permission from the NASA administration for us to build and dedicate the memorials to the astronauts, takes out a glass jar and tucks it in between the feet of the *Inuksuk*. Inside the jar is the agency's official biography of Anderson, one of the HMP patches that we wear, and printed out on a piece of paper the words: "IN MEMORIAM; *Ad Astra Per Aspera;* Erected Summer 2003, NASA Haughton-Mars Project." Then he uses his GPS unit to record a waypoint: N 75 deg 26.005' W 089 deg 52.205'. The handheld device uses the signals from several satellites to triangulate the position of what is, in essence, a prehistoric navigational device.

I scan the horizon to the south, and can just discern our camp and the white hab of the Mars Society. The three architectonic assemblages now define a visually bounded space, the first time that has been done here, the manifestation of an almost inevitable part of the process of humans turning space into place. Likewise, we've been converting the terrain around us into territory by naming prominent features after significant events and people in planetary exploration. We're turning a cognitively difficult environment into scenery, into a staging ground for human activity and history.

Architecture and art, naming and narrative—all are efforts to bring us closer to where we are by making the land ours, always a difficult proposition, if for no other reason than that almost always someone else has been there before us. It's not even a matter of money changing hands and deeds being written, but a cultural presumption that should be negotiated and not taken for granted. In this case, I have qualms about us applying American nomenclature to a land that already carries Inuit names. Joe, his brother Paul, and other Inuit working with us are too polite to say so, but when questioned will shrug and note that their people were here before us, and will be afterward. But naming the landscape can be, I think, as invasive and pernicious an act of assimilation as providing them with DVDs of movies made in Burbank, where I live.

Not far from what has just been named Anderson Pass by the crew is another valley where we ran tests with a concept Mars suit last year, an experiment that in my mind acts as a physical metaphor for the difficulties in grasping the nature of a place. When I arrived on the plane with the suit last summer, it was accompanied by Brian Glass and two guys from Hamilton Sundstrand, the company that has made all of the space suits used by NASA since the Apollo missions in the 1960s. Brian—who is helping us this morning with the *Inuksuk*—is based, like Pascal, at the NASA Ames Research Center. He holds advanced degrees in both geophysics and robotics, and has worked on matters as varied as thermal subsystems for the space station, software for SETI, and maps of the magnetic field of Haughton Crater. He's one of those scientists who seldom reply quickly to a question, but when he does, you realize how focused and driven he is.

Hamilton Sundstrand (HS) is part of United Technologies Corporation (UTC), which owns—well, a lot. It's a $28.2-billion-a-year conglomerate operating in 180 countries with about 155,000 people employed at any one given time, currently making it the eighteenth largest corporation in the United States. In addition to being the parent of Hamilton Sundstrand—which is in itself a $3-billion-a-year company that owns a slew of others—UTC includes Carrier (climate control—or air conditioners to you and me), Otis (elevators and escalators), Pratt and Whitney (aircraft engines for both commercial and military jets), Sikorsky (helicopters), and UTC Power (the only publicly available fuel cells). According to their website, UTC systems are found in more than 90 percent of the world's aircraft. Put together the ability to build a space suit with climate control and fuel cells, and you begin to glimpse the kind of synergy in the military-industrial complex that makes such omnivorous entities succeed. They simply can do some things better than anyone else.

Every decade or so NASA writes a Mars reference mission document, which outlines plans for a future expedition to the planet in just sufficient enough detail for people to develop the next technological concepts for NASA to look at. Then the plan is revised, everyone goes back to the drawing board, and new ideas are again floated. NASA doesn't actually project being able to put people on Mars before 2025 or so, but the idea is to have a planning document that moves equipment design and exploration protocols forward. The current reference mission, set in theory for 2009, is for six people to spend eighteen to twenty months on the surface of the planet, to go on extravehicular activities (EVAs) every other day, and on exploration sorties of up to five hundred kilometers (310 miles) over ten days. The Mars suits have to be designed toward those objectives and Ed Hodgson, the lead engineer pushing the technical design of the Mars suits at Hamilton Sundstrand, sends up a new version to Devon every summer.

The space suits that are used during EVAs outside the International Space Station and the space shuttles are good for only a few tens of hours before they have to be completely overhauled. The current suits weigh three hundred pounds, which on Mars would put 115 pounds on your back. Furthermore, the space suits are more like "wearable spacecraft," as the Sundstrand folks put it, and aren't designed for walking. The Apollo suits worn during moon walks, although offering some mobility, were very tiring to use, and prone to failure from the invasion of lunar dust. Their oxygen supply was good for only about six hours.

The target "felt weight" for a Mars suit is in the fifty- to seventy-pound range (meaning an actual mass between 130 and 185 pounds). It will have to be comfortable, offer mobility over long periods of time, be easy to maintain and repair, be virtually bulletproof, not prone to collecting bacteria on the inside—and did I mention shielding team members from the intense radiation on the surface? Such a suit is not even close to being built, but then, neither is NASA really planning on going to Mars in 2009. What Sundstrand was testing last year and is now is a Hard Upper Torso (HUT) unit for purposes of working out issues with vision and communications. Its white fiberglass shell carries a backpack and has articulated arms ending in bulky gloves.

In preparation for field testing the unit last year, Brian and I had driven out two days earlier with some other camp members to a ridge visible from where we've now erected the *Inuksuk,* but farther to the northwest and out of sight of the camp. There we erected a different kind of monument, the antennae for the "planet phone." A relatively new device, it can skip radio waves along the ground instead of having to bounce them off the ionosphere, which would be a definite advantage on a planet such as Mars, given its paltry atmosphere. We struggled in wind-driven sleet and rain to erect the antennae, but once it was up we were able to use it with a hardened laptop to act as a telecommunication relay device between an ATV parked down in the valley, which we could see, and camp, which we couldn't.

The goal was for Brian to communicate between camp and either a person in the field doing science—or a person mimicking a robotic rover doing science—and with geologists at Ames in a virtual-reality facility. The suit had a camera mounted on it so the controllers could see the terrain in a panoramic view on the screens that surrounded them, enabling them to direct the field presence where to go and what to do. In the case of people acting as robots, Ames would send them instructions about when to move, hold a position, and then move on, all timed to emulate the slowness with which a digitally directed mechanical system would work. Brian had a series of metrics to evaluate the differing rates of return for the time, energy, and money spent on things such as the amount of ground covered, the number of samples collected, and the number of hypotheses tested.

Into this simplified version of an exploration protocol—charmingly designated HORSE for "Human Operated Robotic Science Evaluation"—was injected our resident expert on freshwater lakes and ponds, Darlene. As a geobiologist specializing specifically in paleolimnology, or the history of lakes, she was spending her days wading in the near-freezing waters of Cornell Lake and the ponds nearby. "The ponds here are less than two meters deep," she told me, referring to the many small bodies of water less than six and a half feet deep that freeze solid every year. "The lakes are deeper than two meters and have greater thermal inertia, but their sediments are not as good a record of climate change." She'll spend up to half an hour at a time in hip waders sampling water and sediments, which contain remnants of ancient algal mats and other organic matter, and provide a climate record since the last ice age.

Brian was happy that Darlene had volunteered to wear the HUT for a test, as her background in geology would facilitate communication with Ames. Plus, like many of us, she had worked previously in the Antarctic and was used to extreme environments. We took several ATVs, one carrying the plastic torso inside a box somewhat precariously, and returned to the valley where we had tested the relay. Darlene first put on a telecom headset, and then one that held a tiny cathode-ray tube and a plastic prism that hovered about an inch in front of her right eye. I've worn the latter, and even after practicing to learn how to decouple your eyes so that you can read the screen, it's tiring. But it gave her the ability to hold a specimen up in front of her, have the scientists at base look at it through the camera, then have them relay images with which she could compare it.

The torso unit itself is a challenge to get into, a claustrophobic struggle that takes help from two assistants until it settles on your shoulders. Once the helmet is in place, all you hear are the radio and the whir of the fan cooling you. Pascal still finds it remarkable that you would have to cool a person standing on Mars, where it could be a hundred below zero, but it's a closed system that will cook and asphyxiate you without the fan, a heat dispersal unit, and an air supply. Hamilton Sundstrand invented the requisite technology to keep the suits cool for the Apollo astronauts while walking on the moon.

Once Darlene was locked inside the unit, we spread out behind her so she couldn't see us, isolating her so that she wouldn't pick up any unconscious clues from our behavior, and she headed for the ridge. The guys at camp started her out with a full view of the crater area on her computer, and then zoomed down to a topographical map of the ridge that she would attempt to investigate. The images that were being fed to her were for reference, but also for psychological support, a virtual analog for the multipronged grabbing stick that she was also given.

Used to pick up specimens from the ground, it also gave her a third leg, a useful prosthetic when trying to walk on rough ground with the limited field of vision afforded by the helmet. The view out of the front is good, but your peripheral vision is gone, as is your sense of the sky, both of which we use to orient ourselves in space.

Darlene's assignment was to look for fossilized crinoids and signs of lemmings, as well as to examine the circles and stripes of the patterned ground that showed up in satellite photos. The first clue I had of the cognitive difficulties that she was laboring under was when she didn't recognize that we were walking on frost-heaved patterns until they became so pronounced that you practically had to step over the cracks. Given our propensity to see patterns everywhere, that should give you some idea of how the suit gets in between you and the landscape.

For the next hour and a half Darlene plodded along the weathered limestone ridge, which is just under two hundred feet high and about a mile long. She was directed to climb the south end to ascertain whether or not something green was growing there, but was unable to do so as it was too steep to climb while wearing what I find to be something akin to a plastic bear hug. Whenever I complain about clambering up Fortress Rock in boots and gloves, I remember what it's like to maneuver in the prototype—and that's without the bulky legs and soft bunny boots attached.

Darlene found endolithic bacterial growth as well as crinoid stems along the base of the ridge, and took both photos and samples. She observed moss growing where a lemming had died, and desiccated algal mats in dried-out pools, but was frustrated that she couldn't smell or taste the rocks, which geologists do on a regular basis when in the field—John smelling pieces of Fortress Rock for hydrocarbons is a good example.

When the excursion was over and the Hamilton Sundstrand guys took off the helmet, she was obviously relieved, amazed at how far she had walked, and exclaimed: "Now I know where I am!" She couldn't believe she had covered so much ground.

I talked afterward with John Schutt, who had also had suit time on Devon, and he commented that his depth of field was greatly reduced while wearing it, and that because he's increasingly nearsighted, he couldn't get rocks close enough to his eyes to identify them. Charlie told me that he found the color distortion of the tinted faceplate disturbing, and it made him wonder what would happen to one's color perception after seeing red for a month.

Darlene had been frustrated that she couldn't take notes while in the suit, and this year the suit guys have been watching with interest the work that Bill Clancey has been doing at the Mars Society's new hab in Utah, a somewhat easier desert analog environment in which to work. Bill, a tall and sardonic fifty-one-year-old with a salt-and-pepper beard, is chief scientist for human-centered computing at Ames—but he's also our ethnographer, studying the practices and habits of exploration and seeking to integrate them with robotics and computer software programs. He has a program now called Brahms that allows someone in a suit to make verbal notes while on mock EVAs.

One gray day out at Cornell with Darlene, before she wore the suit, I walked around the shoreline in the drizzle to follow a musical chiming that I couldn't identify. It sounded like pieces of a chandelier tinkling together, and I realized it was coming from a pack of ice that the wind was pushing toward shore. Darlene walked out to the raft of frozen chunks and retrieved a piece for me. "It's candled ice," she explained, handing over a foot-thick fragment as big around as a dinner plate. It was hollowed out through and through, parallel tubes separating long plates shaped like flat candles. The sound it made was the only natural noise I've ever heard on the island apart from wind blowing, waves lapping, a stream flowing, and rocks falling. It was like finding something green growing on a glacier, its music was so unexpected. After watching Darlene and then trying on the suit myself, I wondered how an explorer wearing one would ever discover such a thing on Mars. You can't take notes on something you can't perceive, be it patterned ground or the random notes of frozen water.

After Keith has taken pictures documenting the new six-foot-tall *Inuksuk* and its construction crew—an image that he'll post on his Web site—we remount the ATVs and drive back across the dry lake bed to camp. Once we stow the shotgun and radios and hang up our rain gear in the kitchen tent to dry, I wander over to the orange tent next door to visit Addy Overbeeke, one of the Hamilton Sundstrand suit engineers. He lets me try on the new computer headset, which is even lighter this year. The technology, much of which is store-bought, advances so fast that it's hard for the engineers to keep up. Last year Darlene had to use a wired mouse mounted to her sleeve; this year it's a wireless computer the size of a pocket calculator. It's awkward using a stylus on its screen while wearing gloves, and there are issues with reflected sun glare, but getting rid of wires penetrating the fabric is important to ensuring the integrity of the suit, critical not only for surviving in a low-pressure and poisonous atmosphere, but also for preventing contamination of the Red Planet with human microbes. Given the millions that we carry on our bodies, and the fact that some microbes will inevitably survive the clean rooms and sterility protocols that NASA uses to assemble landers, this may be more of a wish than a reality. Chris McKay, the NASA scientist at Ames who authorized the first HMP foray, estimates that the *Spirit* rover, for example, probably carried with it at least a hundred thousand microbes to Mars. Whether or not they will survive the surface radiation is doubtful, but you never know what might make it while tucked inside a piece of machinery.

Another improvement Addy and others are working on is voice-automated software in the suit. Unlike a voice recognition program, which uses a dictionary of eighty thousand words or so, this one takes only a couple dozen key commands, but it would enable a scientist on an EVA to pull up maps in front of her and to store notes in a dictation program.

"The goal is to make the suit the least labor-intensive as possible," Addy tells me as he adjusts the plastic prism in front of my eye. "Astronauts who come back from doing maintenance on the Hubble Space Telescope get so used to the suit that they say using it is like 'putting on a pair of pants and a shirt.' But a space suit is like a miniature spaceship." While I'm wearing the headset, Addy pulls up a mapping program that lets me touch one point on it and draw a line to another. The screen shows me the distance between the two points, the elevation gained and lost in profile; and the program automatically shades in areas that are visible along the route, versus those that aren't. Addy grins at my reaction. "Most of the stuff that we're testing here is commercially developed and will be on the market soon. If we receive a contract to actually develop a Mars suit, then we'll start building our own, more robust systems."

This year Brian is running more metrics to evaluate robotic versus human performance in the field, and he's putting both experienced people into the suit as well as inexperienced ones. I'm not sure where our visiting videographer, Sam Burbank, falls along that spectrum, but he's one of the people here who have worn both the Hamilton

Sundstrand suits and the mockups used by the Mars Society, which requires the crews staying in their habs to cycle in and out of fake air locks and wear suits while performing chores outside. The Mars Society suits are twenty-three-pound canvas jumpsuits with small backpacks that offer some operational fidelity; that is, they restrict movement and use up EVA time to put on and take off. But the hard fiberglass torso models are actual engineering prototypes. When I ask Sam about it after dinner that night, he reminds me that "the suits used by the Mars Society were designed to offer constraints to exploration. They were fun to put on, but the difference is between driving a go-cart and a BMW. The Sundstrand suit feels more purposeful."

"Costumes," chimes in Charlie, who's sitting nearby. He's also worn both suits. "They restricted vision and mobility, but they ended up being more costumes for media."

I remember last year when John Parnell and I were walking along the Lowell Canal one rainy afternoon. It was the Scottish geologist's first visit and he was poking around the rocks in the streambed, orienting himself to the local terrain. As we came around a bend two people in suits were walking up the stream toward us, one of them gesturing animatedly to the other, who was obviously listening. In front of them and crabbing backward in a crouching position was a guy balancing a camera on his shoulders. The person waving his hands in the air was Robert Zubrin, the founder of the Mars Society, and the other two were a reporter and cameraman from Russia getting some carefully framed footage of Mars on Earth.

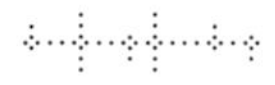

I finish off my day with another evening ascent of Fortress Rock. Sure enough, there's the *Inuksuk* on the far horizon, a tiny protrusion on the edge of vision. Because I personally lifted and carried some of the stones used to make it, I know exactly how large it is, and thus how long it would take me to reach it if I were to drive or walk to it from camp. The first time I walked out of camp to one of the ridges on the horizon, I thought it might take me a half hour. It took twice that. But for all our emphasis on vision's role in our learning about the world, the way in which we know space may depend more on our haptic sense.

Seeing, hearing, smelling, and tasting are all one-way senses. They take in information and give none in return. Only when we use touch do we engage the world bidirectionally; only then do we take in information, then use it to push back, to exchange mechanical energy with external objects, thus changing their shape or position or temperature as we do so. Touch is commonly divided by neurophysiologists into two categories. Tactility, which is based on perceptions gained at the surface of the skin, gives us the sensations of pressure, vibration, temperature, and other physical qualities of objects. Kinesthesia arises from the position of our skeletal joints and from sensations gained through sensors in our muscles and tendons. "Haptic," the root word for which is the Greek *hapteshia,* to lay hold of (or touch), is taken to include both. Haptic perception is a phrase that the Swiss psychologist Jean Piaget first used in 1956 to describe how children formulated space around them, and it is increasingly important to studies of how blind and vision-impaired people sense the world, as well as to the development of electronic gaming and warfare technology, which now rely upon sensitive feedback mechanisms for the control of joysticks, whether the bombs and bullets are real or virtual.

The root of the word "perception" is *percipere,* which in Latin means "to grasp." Proprioception is the ability of the body to keep track of joint positions, which allows us to know, literally, where we are in the world. Along with the vestibular system of the inner ear, it enables us to feel and react to gravity, thus keeping us on our feet. If you lose the sense of touch, all you have left is vision to tell you where your body is in relationship to other objects, which means every move you make must become a conscious act. You have, so to speak, lost your grasp of reality.

Proprioception is an important issue for a space-suit designer. For one thing, the wearer has to receive enough feedback through touch in order to stay on her feet and know where her tools are. Every Hamilton Sundstrand person who has brought a suit component to Devon has remarked on two things when watching a geologist attempt to do work here while wearing it. One, the suit had better be made of a fabric that doesn't rip on sharp rocks. Two, the user had better not hit herself in the faceplate with her geologist's hammer! They are reminded of the latter by the propensity of everyone to run the faceplate smack into the dirt when bending down to pick up something—it's optically clear, so you forget it's there. The grasping stick is designed, in part, to save you from face plants.

That's the most immediate, lowest-level issue. Higher up in the hierarchy of perception is the formation of mental maps. By reacting physically to the land to create a landscape element—the *Inuksuk* that is now anchoring one corner of my vision—I have gained a haptic awareness of the space here. Coupled with the act of vision, my mind now has the ability to do what Addy's computer program did, build me an accurate map of where I am. By adding names to features represented in that map, I organize long-term memory around narrative constructs, and create a landscape. I have wrapped the land around me like a second skin.

The space suit works against this process. When wearing it, you have no direct touch with the world around you. You can read a temperature provided by the heads-up display, but have only a theoretical perception of forty degrees below zero. You smell nothing but the suit and yourself, taste nothing save water carried to your mouth in a tube. You do not feel the wind picking up, and cannot see behind you. So many of your senses are severely mediated that your proprioception is compromised. You brush up against sharp rocks and threaten the integrity of your suit.

When Addy compared a suit to a spaceship, perhaps it would have been more accurate to compare it to a second skin. Ashley Montagu, the twentieth-century anthropologist, noted in his book about skin that the nineteen square feet of the adult male's epidermis acts as a barrier between the organism and the environment. It regulates temperature, blood flow and pressure, stores fat and nutrients, and is a respiratory organ that takes up 12 percent of his weight. All of those are functions in common, more or less, with a suit. The difference is that the skin also contains five million sensory cells, and is so intimately wired to the nervous system that Montagu considered the two to be simply different sides, literally, of the same system. He went on to state that the skin is the oldest and most sensitive of our organs, with touch being the earliest sense to develop in embryos. It is the skin that contains the cells that develop into the mouth, nose, ears, and even into a layer over the cornea of the eye.

In order for a pressure suit to function well enough on Mars to keep people from walking into trouble, or worse, it will have to provide a level of haptic interface with the planet that is far beyond anything even on the drawing boards. The peripersonal space we carry with us—the kinosphere that we call personal space that falls within an arm's reach—is constructed in our minds by mental maps of the world, created by neurons responding to both touch and vision in several areas of the brain, cells that fire when things approach us within that sphere. Peripersonal space responds to a variety of stimuli, including the kinds of clothes we're wearing, the car we're driving, and the tools we use. Darlene's sense of space, by the time she took off the suit and put down the grabber, presumably had been altered to accommodate her larger body and longer reach, but no one knows much about the phenomenon. Any haptic interface system will have to be able to adapt to changes experienced during the day-to-day process of exploration.

NASA Ames has been supporting efforts to apply electronic gaming technology to Mars exploration in the form of a computer program that allows scientists to browse a three-dimensional topographic Mars map with a feedback-sensitive joystick. Aboard the *Mars Global Surveyor* spacecraft—which also carries a high-res camera—is the Mars Orbital Laser Altimeter, or MOLA. Using an infrared laser beam much like radar, it bounces light off the surface of the planet to map it. During 1998 and 1999 it collected twenty-seven million elevation measurements that NASA used to construct a three-dimensional map of the planet. Olivier de Goursac, a French imaging specialist, and his colleague Adrian Lark went one better, and produced actual views as if you were standing at selected points within Valles Marineris. By the time the satellite's instrument stopped functioning as an altimeter in the summer of 2001, it had made something like 640 million measurements of the Martian surface. Now you can do virtual fly-throughs of the canyon and around the volcanoes.

The two scientists at Stanford who came up with the haptic browser used the MOLA map as a visual surface to be translated into texture. They invented a series of algorithms that turned the measurements into a vast array of polygonal object representations. Each map they built—one of Olympus Mons, for example—required them to generate more than one hundred million triangles to create the haptic perception of a single map. That allows you to scan over as many as ten thousand polygons, or contact points, within a half second with your hand while simultaneously watching a cursor follow the contours of the terrain on the monitor. The immersive program gives you some textural sense of the Martian surface through feedback in the joystick, making the map a multisensory tool enabling scientists to identify features of interest. Eventually a derivative of such technology might also allow explorers in suits to feel features in front of them as they approach new terrain.

When Brian sorted out his data from walks last year made by Darlene, John Schutt, and others, he found that humans wearing suits are approximately three to five times more productive than the autonomous rovers that NASA hopes to deploy in 2015. Those rovers will perform thirty to fifty times more work than *Opportunity* and *Spirit*. The difference between a current rover and John out banging a hammer on a rock while wearing his parka? John has a 300 to 400 percent higher rate of return. Haptic features provided to the teleoperator of a rover or a human in a suit could improve the performance of either.

The ultimate equation into which Brian's metrics will be plugged is the one that determines what the mix will be of robotic and human exploration on Mars. An important factor in the equation is cost in terms of safety. How do you balance the value of a human explorer versus the amount of science that can be done by a robot? The most interesting geology (and sometimes microbiology) tends to be on exposed vertical surfaces, places where hiking, mountaineering, and even technical caving and rock climbing skills will be needed. That's far beyond the capabilities of any suit, much less even the most far-fetched rover designs, haptic interface or not.

Ad Astra Per Aspera, the motto Keith placed inside the jar at the feet of the *Inuksuk,* translates roughly as "to stars through endeavor." It acknowledges the difficulty and danger of going into space, of how hard it will be to work on Mars. The Apollo astronauts found that after several hours of working in the suits on the lunar surface, their hands became tired and sore from constantly having to push against the pressure in their suits to move the gloved fingers. Eventually the tips of their fingers were rubbed raw and began to bleed, which is as clear a metaphor for the haptic difficulties as I can imagine.

Learning what the cognitive issues are—learning from people such as Paul Amagoalik and his brother Joe about how to stay oriented when you lose environmental clues—will be critical to the design of the space suit. And you can be sure that one of the functions it will be called upon to perform will be the piling up of rocks into cairns for a variety of memorial purposes.

CHAPTER SEVEN
Mars Air and Ground

PASCAL PUTS the orange Humvee in gear and creeps over the edge of a rocky bench, the triangular tracks on the truck's four axles making it sound as if he's driving a tank. It's early evening, the deeply fractured terrain around us monochromatic in a low light diffused through high clouds that presage a change in the weather. Pascal keeps the boxy vehicle pointed straight down the thirty-degree slope and brakes gently as he starts the descent. Humvees can clamber up and down firm slopes twice as steep as this, but here we're on scree and gravel resting atop permafrost. The front tracks, instead of rolling over the rocks, begin to push up a wave of material ahead of them, and the entire vehicle slews sideways halfway down the hill, tilting dangerously toward the driver's side.

Pascal manages to get the Humvee stopped before it rolls over. The engine idles. The passenger door opens and Charlie Cockell climbs gingerly out, followed by Bill Clancey. Sam Burbank and I are on ATVs working as scouts. We ride down the hill to join them.

After a minute waiting to see if the vehicle is stable, Pascal gets out to survey the situation, then climbs carefully back in to continue down solo. The rocks pile up in front until they completely block his progress. He shifts into reverse. The tracks grab, slip, and then the rear right track suddenly flips backward, ripping into the aluminum of the upper wheel well. Pascal gets out again, looks at the track in disbelief: The equilateral triangle of the track is now standing on one of its points with a second wedged deep into the body of the Humvee.

We gather around to inspect the damage, Sam and I taking off our motorcycle helmets to see better. All of us agree there's a chance that, by driving forward, he might be able to free the unit by rotating it forward. He climbs back into the cab, then slowly revs the Humvee forward. More screaming metal. The Humvee lurches onto a snow patch at the bottom of the ridge, but the track remains stubbornly jammed upward. Everyone yells for him to stop.

Pascal shuts off the engine. Not only is the uphill track screwed up beyond belief, but now the front right track is headed discernibly more to the southeast than its mate on the left. This should be interesting. We're several hours out from camp, edging up into polar bear denning territory, and weather is moving in. We have a shovel, a rudimentary toolbox, and enough food and gear to last us for a few days. This is a less than ideal situation, but it's also part of what we're out here to test: Can we get out of this mess ourselves?

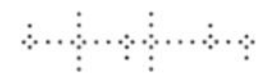

A week earlier Pascal, with another crew, had gotten the Humvee into camp from where it was stranded on the other side of the Endurance River, which they had crossed by waiting a few days until the water level subsided with cooler weather. Once in camp John had reinserted the bolts on the tracks that had been shaken loose—at least the ones that they could find while traversing—and Steve Braham, the camp's communications guru, had installed a radio in the cab. Pascal and John had stocked the vehicle with enough food for a week of travel, the plan being that five of us would head to the coast about ten miles north. The primary goal of the trip would be to allow Charlie and Pascal to examine the water-formed gullies cutting the cliffs above the beach of Thomas Lee Inlet in Bear Bay. These eroded features of Devon's coast hadn't yet been studied, and they appeared to be similar to those photographed by the *Mars Global Surveyor*.

The excursion would also allow Bill to observe and analyze the efficacy of our navigational procedures as we negotiated the isotropic landscape without GPS, a situation analogous to what human explorers might face on Mars. Sam, a lanky filmmaker in his midthirties who has worked previously with both the Mars Society and HMP, would shoot footage for a National Geographic documentary he was putting together about Mars on Earth. And I would, as always, take notes about everyone else taking notes. This would be the first science trip for the Humvee, and between all our efforts, we hoped to learn more about the vehicle's field capability, how the decision-making process would work while en route, and what the tenor of the crew interactions would be in the cab.

We knew that to reach the coast, after examining the aerial photos and based on the drive from Endurance River, we would have to travel up to twice the ten-mile straight-line distance on the ground in order to ford rivers and avoid cliffs. Although we loaded the Humvee with enough food for an extended trip, we figured we could get to the coast in an easy day and a half, look at the cliffs, maybe spend the night there, then get back in a single long day. Our biggest worry was polar bears. As the name of the bay indicates, it's a major denning area, and driving through the valleys would be a deliberate provocation. We planned on sleeping in the back of the Humvee. Its aluminum shell wouldn't keep out a determined bear, but it would at least slow the bear down more than a nylon tent, and give us time to react to its presence.

Three people would ride in the cab at all times, that being the configuration most likely to be used on Mars. Pascal would take turns driving with Charlie, who would navigate using the aerial photos and topo maps—again, as if they were explorers following photos made by the *Global Surveyor* or some other satellite. They would decide on the route, as well as where to stop to examine sites of geological and biological interest. Bill would sit slightly behind and in between them on a jury-rigged seat so that he could observe, video, and photograph the decision-making process, part of his ethnographic studies.

Sam and I would alternate taking point, acting as if we were robots directed by Pascal over the radio to test the routes that appeared most feasible to him. The other ATV rider would ride out in front of the Humvee by only a hundred feet or so as a mud scout. In theory this would allow the driver of the fully loaded 8,800-pound vehicle to gauge whether or not the saturated ground could support the tracks.

Ivan had requested that he and Hernan accompany us in order to shoot the journey. Bill and Pascal decided, however, that they wanted to keep the science and navigational decisions relatively uncontaminated by media. Sam, who specializes in making one-person documentaries in the field, could discreetly take video, but a television crew would, by the nature of the beast, want us to backtrack and maneuver for the sake of good shots. We'd already tried out piloting arrangements with just the five of us a couple of days earlier on the von Braun Planitia, and it was complicated enough without extra people.

Driving a Humvee is itself pretty straightforward. Put it in gear and depress the gas while feathering the brake so that the drive train has a chance to engage the four axles evenly. The high box of the ambulance configuration means not turning quickly. Don't downshift to slow the vehicle, but use the brakes. And with the tracks on, don't go faster on the rocks or snow than around twenty-five miles per hour. Piece of cake. The main difficulty is not allowing yourself to apply too much power to any given situation, lest you break an axle. Or flip a track.

While ambling along on the test drive we came across Larry Young and his NASA Ames crew, who were testing a couple of their unmanned aerial vehicles (UAVs). Flight on Mars is going to be essential in its exploration, if for no other reason than that its surface of 55.6 million square miles equals approximately all of the land surfaces—continents and islands together—on Earth. At the rate that *Opportunity* and *Spirit* are able to explore, it would take millions of rovers to investigate the entire planet within a generation. Aerial surveillance is an obvious solution, and Edgar Rice Burroughs first proposed aircraft on the Red Planet in his 1917 novel *A Princess of Mars*. While he favored dirigible-like vessels powered by antigravity rays, in 1953 Wernher von Braun suggested that more modern, if sadly prosaic, rocket planes be used.

Starting in the late 1970s, NASA began investigating the possibility of using UAVs, and that effort has only gotten stronger as the military adoption of such vehicles has proved successful in other desert environments, such as Afghanistan and Iraq. The aircraft under consideration have included balloons (which drift with the wind), blimps (tethered balloons), dirigibles (powered balloons, in essence), airplanes, helicopters, and biomimetic craft based on insects that flap their wings.

Larry and his crew were hand-launching a custom-made airplane with a small gasoline-powered prop engine and flying it around the playa, sending it from one GPS point to another and watching a wireless video transmission of the ground being covered. Basically their UAV is a high-end version of a hobby aircraft with some advanced electronics, a stepping-stone toward being able to send up vehicles that could fly autonomously to map terrain, to identify features of scientific interest, and to plot routes for ground-based rovers. What they're seeking are workable prosthetics devised to extend our sensory and biomechanical abilities as remote scouts. Our little group would soon find out for ourselves what a boon that would be.

We tried to depart in the Humvee the day after the test drive, but the weather had been foggy, perfect hunting weather for the polar bears, who sport weak eyes but superb noses. In addition, it complicated route-finding and made it difficult for Sam to film, so we held off until late the following day. As we left, Hernan paralleled us just along the airstrip on an ATV to film the departure. We crossed the von Braun Planitia once again, passed over the saddle by the *Inuksuk,* and soon thereafter the valley where Darlene had tested the suit last year. At first we followed a small river up into the highlands, a series of ridges we would have to cross in order to drop into the larger watershed that would take us into the coastal hills.

Pascal had said before leaving: "Driving the Humvee on new terrain, where there's no road, it's a bit like flying—a decision every second." That feeling is an indication of how he operates in the field, a decision-making process that Bill Clancey characterizes as "based on 'least-commitment,' which allows him to make decisions about as few factors as possible at any one time." It's a focused management technique that isn't that far from flying by the seat of your pants, but getting through the highlands turned out to be an excruciating process precisely because our information was deliberately circumscribed. Compasses don't work well this high in the Arctic, so using a GPS unit is standard operating procedure these days for Inuit and NASA alike. But because we were attempting to mimic conditions on Mars, which doesn't have a network of satellites to provide positioning coverage, our use of the GPS was limited to noting waypoints and distances covered. Actual navigation was done by aerial photos and topographical maps, which are very difficult to read in such a dendritic, iterative landscape. The aerial photos showed distinctively shaped snowbanks that had been stable for decades, and these were our primary landmarks.

Pascal drove, Charlie held the papers in his lap, and they radioed Sam and me to "Go toward that hill on your left," or "Turn right and stop," as if we were robots with no initiative. That worked fine through the river valley, but once up in the highlands the terrain was deeply trenched by frost and runoff, the ground covered with sharp dolomite boulders sized from footballs to wide-screened television sets. Bucking the ATVs down into and up out of the jagged maze of patterned ground reminded me of riding through the transition zone of sea ice with Paul several months earlier—only this was rougher. By the time we reached the topmost ridge, the navigation protocols had lost their appeal for ATV riders and Humvee passengers alike. After four hours we'd gone only four miles from camp as the crow would fly, were there any such birds on the island. We'd have welcomed any aerial intelligence.

At that point Pascal sent Sam and me along the ridge in opposite directions to find routes down. One looked longer but less steep, the other steep but a quick route to the river below. It was the last thirty-foot-high bench on the quick-descent option that brought the Humvee to grief.

Now, all of us are standing around the Humvee looking at its torn and crumpled rear flank. A turn signal dangles by a wire. We look at each other, shrug, and break out the tools. The bad news is that we have no idea if we can extricate ourselves from this situation. The good news is that Sam, in between taking stints with a shovel, is getting great footage for National Geographic.

It takes us an hour and a half of digging out the snow, and then the slush from the melting permafrost, to excavate a hole large enough to permit the track some range of motion. Pascal crawls underneath to remove part of the suspension that might be in the way of the track rotating back down. One of the virtues of a Humvee in the field is that you can basically take it apart piece by piece as needed with a set of wrenches and some screwdrivers. We jack up the one side of the vehicle, push rocks into the hole, then let it back down. Pascal pulls himself wearily into the cab, starts the engine, and begins to move forward. The track in back crawls a little farther into the body-work, and then without any warning suddenly flips around into its normal position. We're stunned by the success.

Pascal pulls the Humvee ahead to a patch of rocks, and Sam decides to take a look at the front tracks, which have been forced outward from each other by a few degrees, giving the vehicle a distinctly cockeyed look. Before Sam became an independent documentary filmmaker he was a motorcycle mechanic, and within an hour he's realigned the front tracks using said wrenches and screwdrivers. At Sam's request Bill picks up the camera to film him while he works. Some of the front suspension is bent, but it's designed to handle some serious tweaking. Sam doesn't even have to use the hammer. When he's done, an eyeball check of the alignment doesn't find anything wrong.

We decide to make camp where we are, it's almost ten at night and the sun is behind the ridge. After a quick dinner on ground that gets increasingly soggy, the active layer melting underneath us, we clamber into the back of the Humvee and roll into the bunks of the ambulance body. Pascal gets in last, closes the door, and slots his sleeping pad and bag onto the floor. We go to bed at one in the morning, get up at eight, and have Mars candy bars and bacon for breakfast. Chocolate and fried pork isn't something I'd recommend ingesting first thing in the morning to anyone outside of a polar environment, but we're starved for the high fat content, and the meal warms us up immediately.

After breakfast the first thing we do is decide how long we can afford to go forward. Steve has warned us on the radio that the weather is due to deteriorate in a major way within twenty-four hours. Given the slowness of our progress, we decide to go as far as we can toward the coast and then make a final decision in midafternoon. If we're close enough, we'll push on.

The travel today is nowhere near as rough as yesterday's, but the landscape is unremittingly bleak. We see nothing alive save some lichen, bacterial stains on the rocks in the streams, and every hour or so perhaps a single tiny flower. We mostly follow watercourses, and at one point come within sight of the cliffs above the inlet—but by two thirty we've only come just over half of the driving distance from camp to the sea, and we reluctantly turn back.

It's a difficult decision. Whenever I've been in the field, whether it's with photographers in the deserts of the American Southwest, geologists in the Antarctic, or climbers in the Himalaya, there's always an element of "go fever." It's that push to drive over one more hill, to walk into the next valley, to ascend just a little farther up the ridge. Every single successful exploration has that component, that drive, and we're no exception. It's a feeling made all the stronger by our sighting of the cliffs in the distance. It's not just the vertical scenery that attracts us, but also that we're being promised an elevated view of what lies beyond, the Arctic Ocean. It would put us at the border between land and ice.

But once the decision is made we move quickly. Sam and I reclaim our abilities as autonomous agents, and it takes us only four hours to find a quicker, much smoother way back to camp. We've been taught by the terrain yesterday how to read it—where lighter colored ground will be less broken and faster to travel over, for example—a valuable reminder as to how robots will have to be programmed to learn while on the move if sent out autonomously.

Back in camp everyone is glad to see us and, looking at the bent bodywork, amazed at how well the Humvee recovered from the incident on the hill. We unload the vehicles and rehang the shotguns inside the cook tent. It's both a disappointment and a relief not to have encountered a polar bear. It would be great to see one from a distance, but nature doesn't work according to our wishes, and it's just as well to avoid them. Last year Oz and another geologist had been stalked by one at a lake where they were working. They were forced to fire warning shots and then depart on their ATVs when fog rolled in and they lost visual track of the bear. Nothing like a little fear to dispel any notion of "go fever."

The next morning I spend some time out on the airstrip with Larry Young watching the guys fly their red-and-white radio-controlled plane from one predetermined GPS point to another. The weather is marginal, still threatening, but the storm is not yet here. They circle the plane around Fortress Rock, testing out both a fish-eye camera mounted on the bottom of the fixed-wing craft, as well as a forward-looking one. Ray Demblewski, a skilled model-airplane pilot, holds a control unit in his hands, ready to take over the piloting from the onboard computer if the winds start gusting more heavily. The boss is on all fours on the ground in front of an open waterproof case, staring intently at a Sony portable video viewer receiving wireless video from the cameras.

Larry heads up what's known as the BEES project (bioinspired engineering of exploration systems). It's a project that seeks to overcome the limitations of the rovers by applying foraging rules through low-energy computers in UAVs. Although the round-trip time it takes for communications between Earth and Mars isn't more than half an hour, when you couple that with the fact that the rotation of Mars puts a rover out of antenna reach, and then add to that the decision-making process at JPL (Jet Propulsion Laboratory), the typical result is a single command being sent each solar cycle (a "sol," or Martian day). That makes for very slow rovering. If an airplane could make autonomous decisions to locate features of interest for scientists, the time, energy, and money that would be saved could be enormous. Brian, I have no doubt, would love the metrics.

Last year I'd gone out on a UAV traverse with geologist Emily Lakdawalla from the Planetary Society, a nonprofit organization started by Carl Sagan to promote exploration of the solar system, and one of NASA's numerous private-sector partners. Emily was working with Larry's team to test a remote-controlled commercial hobby craft and also to gain some familiarity with Devon Island as a test site. Joe Amarualik took us far up Rhinoceros Creek where she would have level ground for both takeoff and landing as well as some vertical relief to scan with the camera. One of the virtues of airplanes on Mars is that, while satellites can look only straight down at most features on the planet, airplanes can fly low enough to look sideways at crater walls and gullies, offering an oblique view that brings into focus strata and other geological features. In addition to acting as scouts, in other words, they can collect information directly.

Although Emily and her team were able to receive wireless video from the plane while it was being piloted from the ground within a line of sight, and to record GPS information digitally on ground images, they couldn't get it to fly autonomously using a software map of the terrain and GPS to navigate. This year Larry and his guys are flying predefined flight plans with GPS; furthermore, they set out targets for the cameras to find. When the onboard computer recognized the orange tarp as a feature of interest, a program kicked in that returned the craft to the target, whereupon it automatically released a parachuted probe with a wireless camera directly over it.

What Larry and his people are doing is developing behavioral strategies for foraging based on how organisms locate food and mates. They've looked at the random movements of *E. coli* bacteria as they feed, how moths follow chemical gradients to locate mates, the communal foraging of birds, how bees dance to one another, how sharks locate prey through sensing pressure waves and scent in water, and, at the upper cognitive limit, how foxes and wolves use the landscape itself to determine where prey might be located. The principle being followed is that nature favors survival for those animals that tend to successfully balance the conservation of energy while locating food and mates.

An example used by the team to illustrate this is how foxes search for mice. The small rodents scurry along under vegetation wherever possible, but must venture out into the open to eat. Foxes know that their best chance to find mice is near vegetation, yet need to catch them in the open. The transition zone is, therefore, where they look. Now think of the conflicting roles of flight managers and scientists. The former want rovers to stay in nice, safe, flat areas, but the scientists want to look at vertical features that expose stratigraphy. (Think of *Opportunity* trundling carefully around the edge of Endurance Crater.) Where the two kinds of terrain meet is a transition zone. A BEES vehicle could be programmed to look for such ground, then conduct an onboard survey of interesting features, and send the information to mission planners for future use, or to a team on the ground in a vehicle.

The Hamilton Sundstrand engineers are not on Devon to test an actual pressure suit that could be used on Mars. That stage is years off. They're here to test prototypes so that they can build up an understanding of what fieldwork will require of the suit. The same thing is true for the BEES engineers. Flying a plane on Mars, where the atmosphere at the mean ground elevation is only equivalent to Earth's atmosphere at one hundred thousand feet, will require an extraordinary amount of lift. Designing the wings isn't what the project is about, however; figuring out first what the plane needs to do is the point.

Larry tells me: "We need feature recognition. Someday someone's going to figure out how to fly an aircraft on Mars. We're not interested in that part—we just want to know how to make it useful. The likelihood is that a Mars plane won't be able to fly for a long time or distance, so how do you maximize time aloft? We need to be able to search for what's interesting more efficiently than running a search grid, which takes a long time. So, can we design a craft that can make decisions about following its nose?" Corey Ippolito, one of his young engineers who's working on an emotion-based flight system, puts it to me this way: "How do we develop feature-recognition software that operates on a stochastic or emotional basis, so it spends less time searching than it would if flying back and forth in a grid? Getting mass to Mars is expensive, so an aircraft won't be able to fly for long."

It's probably wise that the BEES team isn't too focused on designing an actual aircraft for flying on Mars, given the crash rate of every plane I've seen flown here. Larry tends to test vehicles first at Moffett Field, where Ames is located, then take his custom-built craft—my favorite this year is the coaxial rotorcraft, a helicopter with rotors both on top of and below its body—and test them next not at a Mars analog site, but at a Haughton Crater analog, a valley in California. Only afterward will he bring them here.

Still, things happen. As the little plane makes another pass at Fortress Rock the soft buzz of the engine quits. Ray takes over its flight with his control box, and brings the plane around in a circle using just the ailerons and rudder. It sets down gently on its two wheels, but then stumbles on a rock and skids to rest on its belly. Ray puts down the controls, walks over, scoops up the lightweight fuselage and examines the undercarriage, which has been ripped off. "The engine probably quit because I was running it at a low throttle and it wasn't generating enough heat to keep it warm and the fuel flowing."

I'm reminded of the wrecked plane with its nose buried in the snow that Paul and I had visited on Cornwallis. Small errors in calculation lead to larger consequences. One of the virtues of UAVs is that there's no human cost involved with a crash. Ray simply carries his plane to camp, where he glues the wheel assembly back onto the bottom of the plane. He'll be flying again tomorrow, if the weather is favorable.

NASA does actually have a candidate vehicle to release in the Martian atmosphere as early as 2008, depending on competition for funds with other projects. It's the cleverly named ARES (aerial regional-scale environmental survey), a craft that looks like a flattened and severely rounded glider. It would be dropped from a spacecraft and deploy a supersonic parachute to slow down, then unfold its twenty-one-foot wingspan and eighteen-foot-long body in three places. Von Braun would appreciate the fact that it would fly for an hour using rockets firing for short bursts every five to ten seconds. After covering more than three hundred miles in that hour flying a mile above the surface, the fuel would be exhausted and the vehicle would crash-land. In addition to measuring the surprisingly strong, if intermittent magnetic

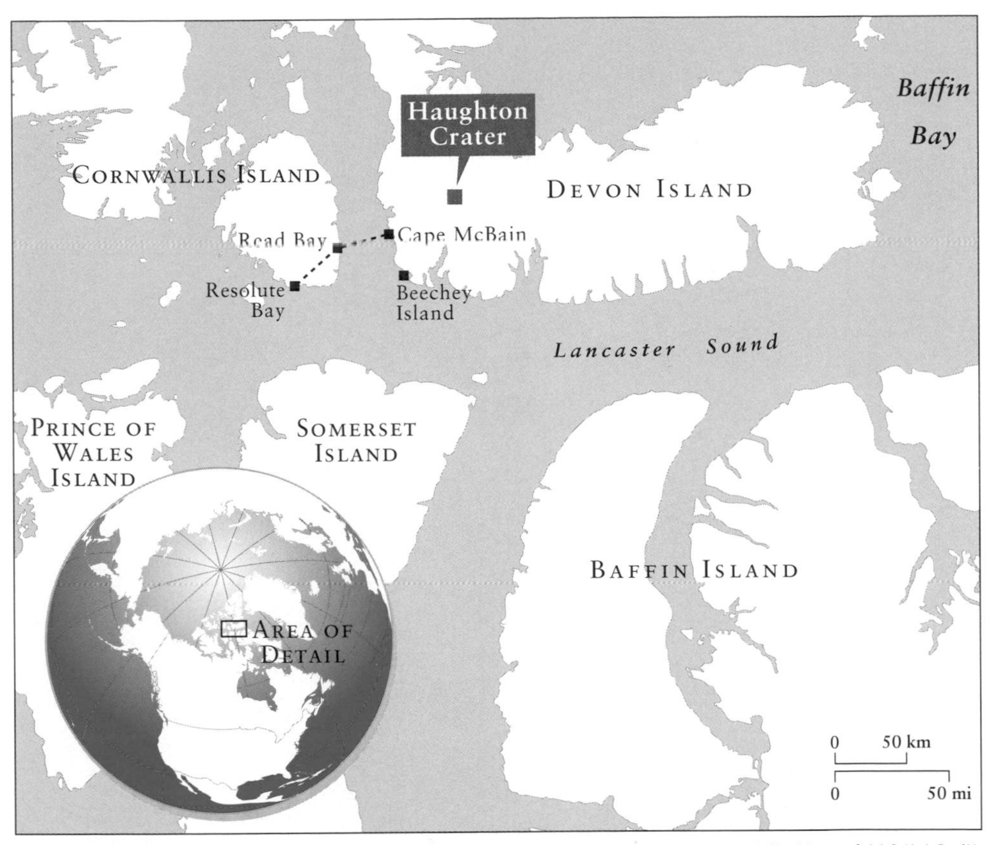

Mike Morgenfeld & Kat Smith

Map of the Canadian High Arctic showing passage of the NASA-HMP Humvee from Resolute Bay to Devon Island. The location of Beechey Island, where Franklin's expedition began to founder, and the Haughton Crater are also noted.

Livio Tornebene

Landsat 7 satellite image of the Haughton Crater. The image, created from near-infrared channels, was taken from 438 miles up in 1999, and has a resolution of 30 meters per pixel.

Peter Essick

An aerial image of the Haughton Crater from 8000 feet

Pascal Lee

Humvee traversing the frozen Wellington Channel en route from Read Bay to Cape McBain

Peter Essick

Valley networks just outside the crater rim

Peter Essick

Detail of a shatter cone

Pascal Lee

Towering cliffs and Mars-like gullies along the north coast of Devon Island. The MARS-1 Humvee rover reached the top edge of these cliffs after a sixteen-hour traverse from camp in August 2003.

Pascal Lee

The HMP camp from a Twin Otter airplane on approach

Pascal Lee

HMP members and the commemorative *Inuksuk*. From left to right: Keith Cowing, Kimmiq, Pascal Lee, Steve Hoffman, Corey Ippolito, William L. Fox, and Joe Amarualik.

Ronald Sidgreaves

Darlene Lim in a Hamilton Sundstrand pressure suit prototype

Pascal Lee

Paul Amagoalik and William L. Fox

Pascal Lee

Humvee with a flipped track

Bill Clancey

Elaine Walker in a music video costume

Pascal Lee

Members of the HMP team reenact Pat Rawling's painting *First Light*

Distant Shores painting by Pat Rawlings

Keith Cowing

The Arthur C. Clarke Greenhouse

Sam Burbank

The Flashline Mars Arctic Research Station

Peter Essick

ATVs in the crater

Pascal Lee

HMP traverse party on ATVs inside the crater

Pascal Lee

Two people on the rim of the Haughton Crater during a simulated extravehicular activity (EVA). Photos shot at the crater often use an orange or red filter to heighten the illusion that the views are taken on Mars.

field of Mars, ARES could provide the first direct measurement of water vapor in the atmosphere.

Most interesting to me is that it is designed to carry a camera and provide the public with an aerial video of the flight. Depending on the region selected for survey, the views could give NASA the kinds of scenic climaxes it needs to keep public interest in, and thus funds flowing for, further efforts. The idea of visuals from down inside Valles Marineris makes most Martian enthusiasts swoon. In addition, John McGowan, writing in Charlie's *Martian Expedition Planning* (2004), estimates that such video could earn NASA up to tens of millions of dollars in advertising alone.

What produces "go fever" is human foraging, a need to look out upon new terrain in quest of territory, to capture space with our gaze and turn it into place. The juvenile polar bears roam across Devon Island because they are pushed out of the denning area by adult males feeling threatened by competition for their already established hunting grounds. Primate groups do the same thing when they reach a population size equal to what their local environment can support. Our urge to explore is not whimsical, but based on millions of years of evolution aimed at distributing growing populations of primates among groups of other animals in a sustainable manner. Four-wheeled vehicles are prosthetic aids to that end, as are the UAVs. They extend our legs and eyes in order for us to arrive more efficiently at new terrain.

Nothing propagates the desire within human populations to experience new places faster and deeper than pictures, vision being the sense that is the primary carrier of information throughout our species. Everyone loves a map, a specialized kind of picture that one person creates so that another can follow. Maps present space in a variety of ways, from the pictorial representations of medieval times and the coastal profiles used by the Portuguese explorers in the 1500s, to the more abstracted topographical maps of the nineteenth and twentieth centuries, and now as printouts of data sets.

In addition, more accessible imagery is used in exploration as a way of raising funds, even by scientific agencies, and the history of Antarctic imagery provides a cultural analog for Mars exploration. Travel and nature photography, guidebooks, paintings and sculpture, and IMAX movies of the Antarctic are all supported with funds from the National Science Foundation (NSF), which administers the U.S. Antarctic Program. The intrinsic merit of such efforts notwithstanding, all of them are also calculated to appeal to the public's sense of adventure, which is based on our genetically hardwired need to explore, and to thus maintain political support for costly operations in a difficult environment.

When James Cook sailed to Tahiti to observe a transit of Venus across the sun, he was also given orders to probe as far south as possible in search of *Terra Australis Incognito,* the "great unknown continent" to the south. By that time the English had adopted the tradition of including visual artists on expeditions, a practice they borrowed from the Dutch, who had learned it even earlier from the Portuguese explorers. These artists performed three important functions: They made sketches that were needed by the military as strategic documents of local topology and native technology; they provided proof to the politicians of where the expeditions had been; and they served as a source of revenue for the artists, who sold them to a public keen to follow their adventures. The images, in turn, created public support for the government to fund more exploration in search of new territory.

Accompanying Cook into Antarctic waters was William Hodges, who went on to become the most widely traveled artist of the seventeenth century, eventually even making it to India with his easel. His paintings helped solidify the English presence on the subcontinent and to establish a fervent orientalism that ran clear through the nineteenth century. His pictorial record, and that of other English artists wandering through Europe after the Napoleonic Wars, spurred the first wave of tourism, initially for nobles, then rich commoners, and eventually the general public.

The inclusion of artists on expeditions has a long history in America as well, most notably in the nineteenth century with the paintings of Thomas Moran and the photographs of William Henry Jackson as they accompanied Ferdinand Hayden on his second exploration of Yellowstone in 1871. Their works helped convince Congress to create our first national park, and were part of an increasing flood of images used by the government to encourage the settlement of the West in order to fulfill what politicians referred to as America's manifest destiny.

Today the NSF maintains an Antarctic visiting artists and writers program to help promote its activities on the continent. The coffee-table book *Antarctica,* which was produced by America's preeminent color photographer, Eliot Porter, after he visited the ice in 1975 with the NSF, is often cited as the primary reason people want to visit the continent. This desire has been increased recently by the release of several movies about Shackleton's *Endurance* expedition. (A corollary effect has been, of course, the naming by NASA of the crater being explored by *Opportunity,* as well as the river crossed by the Humvee on Devon Island.)

NASA, too, administers an art program that has given us images of space travel and other planets before cameras could get there, or of views otherwise unobtainable. The rationale for commissioning artists to imagine the surface of the moon and Mars is founded on what the early explorers knew: To picture a place in the imagination is to create a desire to go there. Once arrived, to "take" its picture is to grasp at least partially the reality of the place and obtain the ability to give it to others. It is a way of turning terrain into territory by mapping and taking emotional ownership of it. A video from an airplane flying across Mars will be even more powerful, of course, than still images. As Bill Clancey observed when we were out on the Humvee traverse and Sam was riding next to it on his ATV with a camera: "Sam knows that a movie is about moving—and he moves even when his subject is moving."

When the Pathfinder mission Web site received nearly six hundred million hits, it was because people were seeking real images of the planet that had just been received and processed. It wasn't quite real time, but it was close enough. If NASA were able to broadcast a video from an airplane flying across the largest canyon in the solar system, it would be one of the most watched events in the history of the world. Cook depended on images by Hodges being reproduced in print by lithographic techniques. Hayden counted on color lithographs for Moran's work, and he gave photo albums of Jackson's to members of Congress holding the purse strings (and, in fact, more money often was spent on printing up the reports of the expeditions in mid-nineteenth-century America than on the journeys themselves). In the past, NSF and NASA have counted on print media such as *National Geographic* magazine, as well as television. Now it's going to be digital.

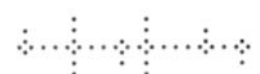

When we were out in the Humvee, we were able to stay in touch with camp via the new radio installed the week before by Steve Braham, a computer scientist from Simon Fraser University in Vancouver who sports an accent and manner of speaking so unique that, when appearing on film, he's often subtitled. I'm not sure I want to probe into how that might have informed his decision to become the director of

PolyLAB, a research center that works on facilitating communications over computer networks—all I know is that when I walk into his tent, he remembers what all of the two dozen–plus computers and servers and hundreds of cables do separately and together. This means we maintain radio and Internet communication with the world and each other. I have faster and better connectivity on Devon Island most days than I do at home. In fact, the megabit-per-second pipeline from our tiny camp on Devon is almost as large as the 1.2 megabits per second (mbps) from McMurdo, the largest base in the Antarctic where more than a thousand people work during the summer.

The amount of information sent back by *Opportunity* is minimal compared to what a streaming video would be from ARES, much less from a swarm of BEES aircraft. McGowan points out that current communication links handle less than one hundred kilobits per second (kbps) between the two planets, and video would require at least one megabit per second. JPL wants a network of communication satellites around the planet, which would also carry GPS, with bit rates from between one and ten megabits per second.

Steve tells me that the current communication pipeline from Mars for the rovers is between 64 kbps (0.064 mbps) to 256 kbps (0.256 mbps), but that an optical system using lasers could carry as much as 32 mbps. "The biggest problem is the light-speed time delay—when one computer will only talk to another when it's told that the other is ready to receive information—that's the real limitation."

Well, that assumes that there's serious money to build such a system, a multibillion-dollar effort that would require a political force fueled by serious public interest, which in turn would rely on the broadcast of compelling images from the planet in a feedback loop that government-sponsored explorers have exploited for centuries. The communications system that we'll need on Mars will be serving many purposes, then—not just facilitating the exploration and science programs, but also keeping the public engaged.

If we're going to send humans to Mars, and they're going to go a-roving, our experience on the ground with the Humvee suggests to us that they will need aerial assistance in scouting science objectives and travel routes, and they will need to be in constant radio and video contact with each other, with remote-controlled rovers, and with mission control. Things will go wrong, constantly, so reliable wireless radio communication will be a safety factor, which may mean ground-based relay stations as well as the satellites. And as Steve notes: "Computers now only communicate with each other in sequence—one sends data only when another computer says it's ready to receive it. That doesn't work with signal delays of up to twenty-two minutes."

Steve also points out that Devon Island is a good analog for some of the challenges facing communications on Mars. "The biggest challenge here is multi-path, which is very much like Mars." He's talking about the fact that radio waves here tend to bounce around and interfere with each other. "First, both places have lots of dry rocks on the ground—the volatiles have been blown out of the breccia, so they are one of the best kinds of rocks we can work on. We're surrounded by valleys and undulating, curved hills that block signals. The way the ridges are around here, there's a natural curvature and it's like being in an echo chamber for radio signals. The signals come in from the field strong but corrupted, which is great. So the shape of the landscape here is really important and better than other analogs, and being in the crater is even better. And then there's the fact that we're sitting on permafrost, which has very specific effects on the propagation of radio waves, and it looks more and more like Mars has permafrost."

All of us use the data pipeline that Steve puts up every year through the satellite dish on the hill behind camp. Personal e-mail is a daily routine, but so are the filing of science and administrative reports, newspaper stories, the teleoperation of cameras and experiments. Keith and Pascal post journals on the Internet for public viewing, while webcams provide the public with real-time pictures of camp life and local weather. The ground and air resources of the camp are promoting navigation on Mars at many levels. In addition to helping explorers and scientists look for sites of interest, then routes to them, they will help the public navigate an alien planet within individual and collective imaginations.

I said earlier that images help people gain an emotional ownership of a space, part of the process of making it a place. You feel attracted to new terrain, all the more so if the views are interesting and even sublime, which is to say both beautiful and dangerous enough at the same time to raise your blood pressure. Those emotions are strongest if you're there in person, but they can also be evoked through televised images and art. The ultimate responsibility of the image makers and takers goes beyond the navigation of terrain. It enters the philosophical and political realms where getting beyond the notion of "ownership" is essential, and Mars is understood as an environment that needs protecting for both its utilitarian and intrinsic values.

Views of a landscape from an elevated vantage point tend to abstract the features because small details are lost. The higher the view, the more it devolves toward being a map, which is (among other things) an abstract of a view or picture. Traditionally, you make a topographical map by looking at aerial photographs; now the process is even more distanced as it is based on data sets produced by remote sensing instruments, such as radar and laser beams.

When we view the world—any world—from above, we are led unconsciously to believe we exercise some measure of control over it. In part that assumption derives from the physical fact that it takes energy to obtain an elevated view, whether you climb a mountain or boost a rocket into orbit. And what we expend energy on we tend to value, thus want to own. Further, as we extend a rationalized cartographical grid over the surface below us, we believe that we have come to understand it because we have measured it. And, to some extent, we have. But there is an ineluctable arrogance that comes with the literal position of being above your subject, an attitude of which you have to remain conscious in order to counter it (the sexual metaphor that identifies a planetary body with the body of a person is an old and noteworthy trope). The arrogance can make us forget what we do not know about the surface, which in sum is far greater than the knowledge we gain by mapping.

The views we have of Mars from above, whether from satellite or in its thin, thin air, will need to be countered by views from the ground, where we are not removed from the environment but part of it. The more personal we make those views through art, the better chance we will have for a balanced view of the planet. We won't just feel that science and technology allow us to own Mars, but that they bring a responsibility for stewardship.

CHAPTER EIGHT
The Face of Mars

WHEN THE STORM hits camp the next day, it whips up seventy-mile-per-hour gusts, not far short of hurricane level, which starts when sustained winds hit seventy-five miles per hour. Sam and I and others stagger over to the tent city to add rocks to the guy lines. Every year someone loses a tent—ripped walls, snapped poles, gear scattered. This time it looks like it's going to be Pascal's tent, the most weather-beaten and frayed of the bunch. Despite the violent flapping of its coated nylon rain fly, accompanied by ominous ripping sounds, the body of the tent survives the blow, and his gear stays dry. The storm moves through quickly, and when the skies clear Sam takes the opportunity to lead Elaine Walker carefully up Fortress Rock, holding his videocam in one hand and helping her balance on the unstable dolomite blocks with the other. A light breeze is blowing and the temperature is above freezing under a mostly blue sky. Sam is wearing a parka and his usual multi-pocketed camp pants over long underwear, typical summer Arctic garb. Elaine is costumed in silver gray tights, a short, bright red skirt, red and gray gauntlets, a red bodice, and a clear bubble helmet. Not typical Arctic fare.

Inside her helmet—to which Addy Overbeeke, ever the practical engineer, has thoughtfully added air holes lest she suffocate, a detail she'd forgotten—she is wearing a short platinum wig that covers her brown hair. She and Sam are climbing the crag so he can get some "pretty shots" to insert into the music video that he's making for Elaine's new song, "We Can Live Here," which is about colonizing Mars. When they reach the top he positions her at the northern prow, the von Braun Planitia a spacious background for lyrics about butterflies and trees on Mars.

It's been humorous and instructive this week to watch him work with the petite pop singer as she lip-synchs on the hood of the Humvee with Pascal behind the wheel, or poses next to the Arthur C. Clarke Greenhouse, a ribbed plastic structure that looks just futuristic enough to stand in for a habitat on Mars. Sam, whose face looks like it should be in front of the camera instead of behind it, makes documentaries in the field that are slightly personal, more like journal entries than formal narratives. Most of the time on Devon he's been shooting science projects and the landscape for the National Geographic feature he's putting together, which requires a relatively sober visual rhetoric and a level horizon line. He's happy to work with Elaine, tilting the camera this way and that and allowing his own aesthetics some increased play.

Elaine, who is visiting the camp in order to work on the music video as an educational outreach activity, has two mathematicians for parents, and earned her first music degree from New Mexico State University in Las Cruces. In 1991 she earned a second degree from the prestigious Berklee College of Music in music synthesis production, and in 2001 a master's in music technology from New York University. Her thesis at NYU was "Music Composition and Building MIDI Controller Based on Chaos Theory." When she's not serving as the president of the New York City chapter of the National Space Society, or writing, performing, and producing songs for her band ZIA, she edits television cartoon music for a living. We share a table in the cook tent where we have our laptops set up next to the large propane heater—a necessity for typing—and while I write, she patches together music clips and audio pieces for episodes of the popular *Pokémon* series, which she e-mails off each day.

"I had this 'overnight conversion' to being a space activist. I was just obsessed with fear of the human race disappearing from Earth . . . dying off," she told me while we were taking a break one chilly afternoon. "After I graduated from Berklee I joined a bunch of online newsgroups about science. I was an ardent environmentalist, but using technology in my work and interested in futurism. When I found out that most enviros were anti-tech, it really dismayed me.

"What we want to do is know everything. There are lots of parallels with religion—getting closer to God, which is the universe, means knowing everything. And to do that we need more technology, which makes us more godlike. Most fundamentalist religions would find that offensive, but I don't understand why we have to rely on these old stories, or make up anything. We could have a dynamic religion where everything was always changing.

"I was a member of the National Space Society, and I remember when Robert Zubrin broke off from that to form the Mars Society. He was such an inspiring speaker. I went to the Mars Week at MIT in 1999 and met Pascal, and as I learned more, I realized Mars is a place we need to go. People need to explore space because we're hardwired to explore. We've been this way since time immemorial, and those cultures that explore uncomfortable territory are those that advance most rapidly.

"I started writing music with my band ZIA in 1992, and when I joined the NSS, I made the decision to pretty much devote my life to writing music about humans in space—the movement. Now I have more connections in the space and aerospace communities than I do in the music world." Like Pascal, Sam, Bill Clancey, and others in camp, Elaine was involved with the Mars Society until recently; when she talks about the human urge to explore, and how colonizing new places is tied to cultural vigor, she's essentially paraphrasing its goals.

Sam, who is a bit more critical of that stance, also refers to the fervent desire to go to Mars as "the movement," although he got to Devon Island by a far different route than Elaine. He was already interested in student films and was writing screenplays when he dropped out of college in the mideighties to run a motorcycle shop at Sears Point, the motor racing complex thirty minutes north of San Francisco. In 1993 he started working as a technician for movie productions, then doing lighting for commercials. Soon thereafter he was shooting documentary work. He's also a musician in the infamous Extremophiles band, which is made up of various Mars scientists and aficionados, and he now maintains a small recording and film studio in the city at the decommissioned Hunter's Point Naval Shipyard. During the way back to camp on the Humvee traverse, Sam and I had stopped for a break when we'd gotten ahead on our ATVs, and I took the chance to ask him about how he got involved with Mars.

"When the Mars Society got started in 1998, I was in a position to offer something. I could be a capable mechanic, a documentary filmmaker. It's how you get to Mars without a PhD. I always loved space as a kid—I watched *Star Wars* in the theater during its first run twenty-five times, and I related to both Luke Skywalker and Hans Solo. In 1999 I was just enraptured with Mars. I'd been reading Kim Stanley Robinson's trilogy about Mars, and I think it changed my chemistry. When I was here in the Mars Society habitat in 2001, I was constantly framing Devon Island as Mars. But I got burned out by the infighting among the Mars groups, and I literally can't put on Mars-colored glasses anymore." Sam had paused then, taking time to look out over the barren landscape. He shook his head.

"There was this sense when the Mars Society and HMP joined forces that 'We can do this!' Now we're back to the long haul. The reaction to September 11 changed the context of the Mars movement. It allowed the president to steer the country into a giant budget deficit. When there was this mountain of black ink, the surplus, there was this 'Let's do something with this that's worthy of a culture on the upswing.' The world doesn't look like that anymore. It doesn't change that we're going to go there, but it changes the how and when. I feel more like an interested journalist now than the 'first man on Mars.'"

As I watch Sam and Elaine maneuver gingerly on the narrow top of Fortress Rock, I can't help but think of the science-fiction movies that came before *Star Wars,* the ones made in the 1950s and '60s that set in motion the careers of so many scientists at NASA and JPL—pictures such as *Conquest of Space* in 1955. It was an era when science and technology were catching up to a strain of fiction that had a century of wishful thinking behind it, a powerful synergy.

To understand why and how Sam is filming both a National Geographic documentary and a music video about Mars on Devon Island, you have to know something about how Mars has been depicted visually in our culture. And to do that you need to start with how we've pictured the moon, because that is the nearest body in space to Earth, the first to receive careful observation—and, in fact, the only off-planet surface we can see clearly from our own through even a telescope. Look at Mars through the largest optical telescopes on Earth, as has been done with the two-hundred-inch reflector atop Mount Palomar, and all you get is a colored blur. Even the Hubble Space Telescope can't do much better. But look at the moon through even a modest telescope in your front yard, and you feel as if you're standing on its surface.

Among the earliest recorded observations of the moon by a Westerner were those made by the first-century Greek writer and thinker Plutarch (46–120 AD), who was one of two priests serving the temple at Delphi and interpreting its oracle. He compared the appearance of the moon to terrestrial mountains and seas, noted that its lack of clouds might mean it was a desert, and speculated about lunar life. During medieval times, observers tended toward figurative comparisons, such as when Albertus Magnus identified lunar features as looking like dragons in the thirteenth century. With Leonardo da Vinci, who sketched the moon without interpreting it, Europe marked a return to more literal attempts at depiction. When Galileo made his first telescopic observations in 1609, he echoed Plutarch in saying that the moon was covered with high mountains and deep valleys.

Prior to completing their 1840 map of Mars, the two Germans Beer and Mädler spent nine years, from 1828 through 1837, compiling the most detailed picture of the moon to date, a beautifully hachured representation of its topography that they included in a four-hundred-page book presenting everything known about the planet to date. At the same time, Franz von Paula Gruithuisen was insisting that he had seen a lunar city composed of lines, which he sketched to resemble a capital letter "A" with multiple crossbars. Looking at a photo of the same region today, we can see how his mind connected together features that were discrete, the same kind of illusion that would plague Lowell at the end of the century.

Although daguerreotypes were being made of the moon in 1840, just months after the technology became publicly available, the most detailed lunar photography made during the century wasn't of the moon itself, but of models constructed by James Nasmyth and James Carpenter, the images of which they published in their 1874 book *The Moon*. Nasmyth was a famous engineer—son of the painter widely considered to be the father of Scottish landscape art—and a serious amateur astronomer. His friend Carpenter shared Nasmyth's passion for astronomy. Their plaster models, most of them only one- or two-feet-square in extent, along with the heliotypes they took of them, set the tone for the next century of both lunar and Martian landscape representations.

They started by enlarging a copy of the Beer and Mädler map from its original diameter of thirty-seven inches to six feet, then shading it in so that it approximated what the moon's surface would look like if illuminated all at the same time from a single direction. The view was a physical impossibility, but one that created a comprehensible picture of the moon, versus a map. Working from that enormous image they made detailed models of its more impressive features, working from their thirty years of close observations and pencil sketches. Then they carried their models outdoors and photographed them in the slanting afternoon light. Their photographs were so realistic that they were then, and are still now, more often than not mistaken at first glance for pictures made directly of the surface. I have a page from *The Moon* with their picture of the Triesnecker Crater sitting on my desk at home, and even with a magnifying glass I cannot tell it is a photograph of a plaster model.

Just as earlier ages had purported to observe on the lunar surface features that supported theories current at the time about the nature of the universe, Nasmyth and Carpenter applied the newly minted theories of terrestrial volcanism to the moon. They had visited Vesuvius during a small eruption in 1865, and observed landforms in the crater that seemed analogous to lunar ones, which led them to make statements such as: "The geological phenomena of the earth affords unmistakable evidence of its original fluid or molten condition, and the appearance of the moon is as unmistakably that of a body once in an igneous or molten state." While they correctly postulated that neither an atmosphere nor water existed on the moon, they tended to overestimate the height of mountains and describe the surface as if they were travelers going on the equivalent of a grand tour.

> The mountains . . . are collected into groups, but mostly they are connected into stupendous chains. . . . The scenery which they would present, could we get any other than the "bird's eye view" to which we are confined, must be imposing in the extreme, far exceeding in sublime grandeur anything that the Alps or the Himalaya offer; for while on the one hand the lunar mountains equal those of the earth in altitude, the absence of an atmosphere, and consequently of the effects produced thereby, must give rise to alternations of dazzling light and black depths of shade combining to form panoramas of wild scenery that, for want of a parallel on earth, we may call unearthly.

They described the most prominent mountains on the moon, the Lunar Apennines, as follows:

> The upper portion of our illustration is occupied by the magnificent range of volcanic mountains named after our Apennines, extending to a length of upwards of 450 miles. This mountain group rises gradually from a comparatively level surface towards the south-west, in the form of innumerable comparatively small mountains of exudation, which increase in number and altitude towards the north-east, where they culminate and suddenly terminate in a sublime range of peaks, whose altitude and rugged aspect must form one of the most terribly grand and romantic scenes which imagination can conceive. The north-east face of the range terminates abruptly in an almost vertical precipitous

> face, and over the plain beneath intense black steeple or spire-like shadows are cast . . . mountains . . . rise to a height of 18,000 to 20,000 feet.

In addition to the fact that their text was a serious effort to understand the morphology of the moon based on a presentation of scientific theories (which even if wrong gave the book credibility with readers), the realistic pictures added a veracity that was otherwise unavailable to other publications. When they included oblique views of the moon, such as "Group of lunar mountains, ideal lunar landscape"—which shows small craters on a plain with towering, steep-sided Himalayan-like peaks in the background—the viewer would have a hard time arguing with their representations.

Their language expresses a corollary to the search for sublime terrestrial scenery that European explorers and scientists, and then artists and tourists, had undertaken for the past two centuries. To vastly oversimplify the process, the explorers had run out of all the easy places to travel across and map, and the scientists found that dramatic landscape features such as large mountains, deep canyons, and desert islands revealed evidence of everything from geomorphological process to the evolution of animals. The artists accompanying the expeditions brought home stunning images of scenic climaxes that tourists then wished to see for themselves, all of which in turn helped to foster the romantic movement in landscape art. Among other attributes, such paintings featured humankind as a small and plaintive presence in the face of a supposedly pure, apparently indifferent, and definitely overwhelming nature.

Even exploration artists, who were charged with capturing objective views of newly discovered lands for their governmental patrons and an eager public, fell prey to the desire for romanticized landscapes, and thus tended to exaggerate the verticality of the terrain. The Arctic images brought back by Elisha Kent Kane from his failed attempt to find Franklin are a prime example. Kane's account of the 1853–1855 journey was the most popular book of its time in America next to the Bible. It contained images of impossibly tall icebergs towering over his ship, in essence the sublime scenery of the Alps translated into ice as a metaphor for the very real dangers facing his crew. F. W. von Egloffstein, accompanying the Colorado River Exploring Expedition during 1857 and 1858, did much the same thing with the Grand Canyon, narrowing and heightening the entrance to Black Canyon until it appears as if it is a chasm entering the netherworld.

Nasmyth and Carpenter weren't, however, blind to the effects of various optical illusions and distortions, and explained in a footnote:

> In reference to such prominences on the lunar surface as cast steeple-like shadows, it is well to remark that we must not in all cases infer, from the acute spire-like form of the shadow, that the object which casts the shadow is of a similar sharp or spire-like form, which the first impression would naturally lead us to suppose. A comparatively blunt or rounded eminence will project a long and pointed shadow when the rays of light fall on the object at a low angle, and especially so when the shadow is projected on a convex surface.

Nonetheless, the depiction of the moon as a romantic volcanic landscape of jagged peaks would dominate the cover of every science-fiction novel about the moon, and every movie set upon its surface, until the *Apollo 17* astronauts landed at one end of the Apennines in 1972 and took firsthand photographs of them. The mountains are, indeed, huge, rising fifteen thousand feet above the plains—but the topography is rounded, blunt, covered in dust, and not at all sublime in the conventional nineteenth-century manner.

Science fiction art, which is defined as imaginative art done to illustrate literature, shares a foundation in common with its more realistic cousin, space art. Their joint origin goes back at least to the illustrations of Emile Bayard and A. de Neuville for the Jules Verne novel *From the Earth to the Moon* published in 1865. Space art, which seeks to depict as accurately as possible alien worlds and travel in space, relies upon information from scientists and engineers. Verne, although writing fiction, was scrupulous about keeping current with scientific progress, and even commissioned a map of the moon from Beer and Mädler for his reference.

The Nasmyth and Carpenter images are cited by the International Association of Astronomical Artists to be the first examples of space art, but it is the boundary between the two, the transition zone from fiction to fact, if you will, where we find the cognitive difficulties of encountering Mars exposed and available for analysis. The visual representation of spaces foreign to us—whether it is the Sahara or the Antarctic on Earth, the Lunar Apennines, or the Valles Marineris on Mars—seeks vertical relief to provide scale and definition. It is another visual mechanism, like that of building contours from adjacent dots, to assemble a picture we can understand. And it is one prone to manipulation, unconscious or otherwise.

The first artist to devote himself to depicting other planets was Lucien Rudaux (1874–1947), a commercial illustrator who built his own observatory and made genuine contributions to astronomy. Although Rudaux's work is well known in space art circles, ironically it was his own insistence on accuracy that kept him from becoming as popular as later practitioners such as Chesley Bonestell. Rudaux observed the lunar limb—the curved edge of the moon—and saw that the mountain shadows were not always the sharp points depicted by Nasmyth and Carpenter, but rounded, the peaks having had their ridges smoothed by millions of years of micrometeoritic impacts and dust. Likewise, when he painted Mars he chose to eschew the idea of canals and scenic climaxes, and instead show flat patterned desert ground with thin clouds of dust in the atmosphere, views not unlike those sent back by the rovers.

Chesley Bonestell (1888–1986) was the most influential space artist of the twentieth century, and while he knew Rudaux and admired his work, he didn't think it was dramatic enough, or even realistic. Trained as an architectural artist, Bonestell also maintained a slow-burning passion for astronomy that had started when, as a seventeen-year-old, he had a chance to view Saturn at the Lick Observatory. After Bonestell turned fifty he started a second career as a matte painter for movie backgrounds at RKO Pictures in Hollywood. In 1944 he sent *Life* an unsolicited painting, *Saturn as Seen from Titan,* which shows the great ringed planet hanging vertically in front of the viewer, the rings seen on-edge as a thin line. The matte work had improved his handling of oils, and the almost photorealistic appearance of the work made him an instant celebrity. As his biographer Ron Miller put it, it was as if readers of the magazine had received a snapshot from *National Geographic* made on another planet. And, in fact, Bonestell consciously adopted imaginary camera angles as his composition technique when depicting the planets as seen from the surfaces of their moons. Miller points out that it makes you feel as if you're on a tour of the solar system, and that numerous people have credited Bonestell's images as the inspiration for them to become engineers and scientists devoted to space exploration.

While working in London in the 1930s, Bonestell had met both Rudaux and another astronomer-*cum*-space artist, Scriven Bolton, an Englishman who also used plaster models to aid his imagination.

Bonestell was a great admirer of Maxfield Parrish and Gustave Doré, and while Bonestell, too, learned how to make models of lunar landscapes, he also adopted the glazes and romanticism of Parrish and the gothic landscapes of Doré. His hard-edged, crisp style was true to the science of the time, but the vision of Mars held by both artists and scientists was still mired in the romanticism of the previous century.

His illusions of standing on extraterrestrial surfaces were just that, illusions. He insisted on portraying the lunar mountain as had Nasmyth and Carpenter. His needle-nosed spaceships are set down among the shadows of jagged peaks, not the rounded ones of reality. Space pioneers Robert H. Goddard and von Braun were launching rockets up to the edge of space, science fiction writers were adopting more hard science over operatic fantasy in their novels, and a hyperrealistic graphic style appealed to the public's increasing desire for supposedly objective depictions—so science fiction illustrators, admiring both Bonestell's skills and his popularity, followed his lead. Even most of the leading space artists today—whose more accurate depictions hang in the Smithsonian and NASA art collections, and who design alien worlds for Hollywood—acknowledge their debt to Bonestell.

The Saturn painting brought Bonestell an introduction to the science writer and novelist Willy Ley, and he was commissioned to illustrate Ley's 1949 book *The Conquest of Space,* a nonfiction book that featured fifty-eight paintings by the artist. Bonestell collaborated with Willy Ley and Wernher von Braun on another book, *The Exploration of Mars,* which was published in 1956 and subsequently made into a movie (with the inadvertently confusing title of *Conquest of Space*). Once again, Bonestell's illustrations of Mars followed earlier conventions, showing it crossed by canals bordered by vegetation, an idea still vigorously promoted by Ley and von Braun. One of his most well-known images from the book makes Mars look somewhat like southern Utah: high mesas and bluffs sculpted in orange rock backdrop a spaceship that stands vertically on its fins. Ironically, the movie portrayed the planet more accurately, showing an eroded and relatively flat landscape—but Bonestell complained that it was not at all realistic.

Bonestell insisted on characterizing the lunar and Martian surfaces in a romantically enhanced manner even after it become obvious from the images televised back from the Ranger missions in 1964 and 1965 that the moon had a much less dramatic surface. Albert Bierstadt had done much the same when painting his late nineteenth-century views of the American West. His enormous canvases continued to feature mountain ranges towering improbably over severely foreshortened landscapes, even as photographs were demonstrating that places such as Yosemite and the Rocky Mountains, although striking enough to be classified as sublime, were much less visually compressed and dramatic in reality. Bierstadt was playing out the endgame of romantic art in support of the American dream of manifest destiny, providing imagery that glorified the new territory first to would-be emigrants, and then later to tourists. Bonestell was supporting the same goal. Despite his hard-edged, hyperrealisitic style, he romanticized a new frontier in hopes of encouraging its colonization.

If the inclusion of a strong vertical element is the most frequently deployed element to anchor a landscape composition (whether it is a tree in the foreground or a mountain in the background), then the use of a human figure is the traditional corollary device used to establish scale. Most traditional depictions of Mars include a spacecraft and/or astronauts, and already in 1945 Bonestell had incorporated spaceships into his work. The NASA Art Collection, started in 1963 and now numbering more than eight hundred pieces, has commissioned a variety of notable mainstream figures from the art world to depict major events in the space program. While the artists have varied in style from the traditional pictorialism of Norman Rockwell and James Wyeth to more contemporary modes of representation practiced by Robert Rauschenberg, the Starn twins, and Vija Celmins, their works for the most part focus on people, a way of placing the public in the action, of providing a literally human dimension to space. The agency curators over the years have acknowledged their debt to nineteenth-century American exploration artists, citing the example of John Wesley Powell hiring Thomas Moran to paint the Grand Canyon.

The agency's Advanced Space Exploration Art collection, which deals with Mars and other planets, features works commissioned for calendars, posters, and publications, and is much more centered around the imagined depiction of alien planetscapes. Its artists tend to be "hard space" practitioners, people who create graphic art that hews closely to technical accuracy, as if to offset the necessarily higher imaginative quotient in the content of their work. NASA maintains a careful distance from these depictions, however, as indicated by the following disclaimer from its Web site:

> Note: NASA currently has no formal plans for a human expedition to Mars or the Moon. This image and others displayed may not reflect the hardware and overall concept of possible visits to either of those celestial bodies. However, the artwork represented here serves as a comprehensive study of various concepts and ideas developed as possibilities over a period of years. The renderings were accomplished by NASA and/or NASA-commissioned artists.

One artist with a large number of paintings in the exploration collection is Pat Rawlings. Born in 1955, he is an heir to the Bonestell realist tradition who specializes in space vehicles and bases. Perhaps the most famous Rawlings image is *First Light (Exploration of the Noctis Labyrinthus canyon system on Mars)*, which shows two explorers on the Red Planet. Following in the model-building tradition of Nasmyth and Carpenter, Bolton, and on through Bonestell, Rawlings built a relief map based on Viking orbiter data and photographed it with a wide-angle lens while it was lit by a slide projector used to represent the sun. He assembled the images into a ninety-degree panorama that became the background for the painting. In the right foreground stands an astronaut atop a red formation that resembles Fortress Rock in shape and size; a second figure in a pressure suit rappels down the cliff to the valley floor.

Although the landscape and foggy atmospheric conditions of the painting hew closely to NASA's photographic images, the idea of an astronaut rappelling is far outside the conservative EVA protocols favored by the agency, and this caused some controversy when the painting was first shown. It's an irresistibly heroic action scene, however, and Pascal liked the image so much that he actually had his brother, Marco, belay him atop Fortress Rock in 2001 as he rappelled down it for a photographic restaging of the painting, which was in itself an imagined re-creation of an event that has yet to happen in the future. (On the left-hand side of the photo stands a small *Inuksuk* atop the rock, a mute contrast to the futuristic action.)

Most of us in camp grew up with paintings of Mars on the cover of science fiction novels and in the background of movies, and Pascal is no exception. He favors the work of Rudaux, Bonestell, and Ludek Pesek, a Czech-born artist who immigrated to Switzerland and depicted the planets more in the style of Rudaux than the dramatic Bonestell. Pesek was an illustrator and artist who worked on major science exhibitions and was commissioned by *National Geographic* throughout the 1970s and 1980s to create paintings for the magazine to illustrate NASA's various planetary missions. A particular challenge for him was to do the paintings before seeing the actual images so that the magazine could publish the illustrations concurrent with the missions, and he had remarkable success in forecasting accurately the surface of Mars.

This business of imagining future moments in Mars exploration has been extended even further by Charlie Cockell, whose small nonprofit, Earth and Space Foundation, has commissioned the largest contemporary collection of Mars exploration paintings in private hands. Charlie notes with undisguised glee that Mars offers not only a mountain two-and-a-half times taller than Mount Everest, but that its north polar ice cap is as large as Antarctica and some of its deserts larger than the Sahara. The dozen or so Mars paintings that hang on the walls of his small house in Cambridge, England, include works by Rawlings, Michael Carroll, Andrew Stewart, David Hardy, and Marilyn Flynn. An oil titled *Conquest of Olympus* by Flynn in 2002 shows a figure on the summit of the giant volcano, arms upraised to hold a flag aloft, a pose Flynn adapted from the picture of Sir Edmund Hillary taken by Tenzing Norgay when they became the first people to ascend Mount Everest. The astronaut in the foreground wears a patch on her shoulder that displays the logo of the Earth and Space Foundation. (The collection can be found online at the foundation Web site: www.earthandspace.org.)

This looping self-referential world of Mars imagery is a nested puzzle of inside jokes, psychological reinforcement for true believers in Mars exploration, and public relations meant to bolster efforts by NASA and other agencies for missions to the planet. As Ray Bradbury once declaimed, we start with romance and build toward reality, and Mars will already host a ghost culture when we arrive there. Elaine, like Sam and Charlie, is one of the true believers in that process, hence her balancing precariously on the top of Fortress Rock while Sam records her lip-synching lyrics. The image of her singing atop a desert cliff will be inserted into a chain of imagined moments about Mars where, as Kim Stanley Robinson notes in the second novel of his trilogy, *Green Mars,* "noosphere preceded biosphere—the layer of thought first enwrapping the silent planet from afar, inhabiting it with stories and plans and dreams."

A relatively recent occurrence of Martian fabulation shows how human cognition can still morph a remotely-sensed image into something that doesn't really exist, and manifest a wish-fulfillment. The face of Mars became the "face on Mars" in July 1976, when the *Viking 1* lander was photographing possible landing sites for the *Viking 2* lander. One of the pictures it transmitted back to Earth was the image of an eight-hundred-foot-high butte in the plains of Cydonia east of Chryse. The hill was situated along the escarpment that separates the cratered highlands to the south from the lower and smoother lowlands to the north, and seemed to resemble a helmeted face. The formation was over a mile wide and two miles long, and although its left side was cast in deep shadow, its right was strongly lit and showed what appeared to be an eye cavity, as well as a strongly profiled nose, mouth, and chin. It all looked vaguely pharaonic. NASA proclaimed the image to be simply a coincidence, but a rapidly growing group of conspiracy theorists insisted that it was a monument built by a superior extraterrestrial race, and that NASA was suppressing what it knew about the origins of the formation.

The furor over the image was fed by a self-educated pseudoscientist from New Jersey, Richard C. Hoagland, who promoted the conspiracy theory over Art Bell's late-night talk radio show broadcast from Nevada, and then on his own Web site. He produced enough noise that NASA instructed Mike Malin—who built the camera aboard the *Mars Global Surveyor,* and whose company, Space Science Systems, processes the thousands of images the *Surveyor* continues to give us of Mars—to take a better picture of the feature as soon as possible after its deployment in September 1997. In April 1998 Malin's camera imaged the hill with ten times the resolution of the first picture; it appeared that the "face" was simply a naturally eroded landform. It was winter at that latitude on Mars, however, and clouds obscured most of the right side of the butte. Hoagland and others insisted that haze was obscuring the true details, so Malin and the mission control people at JPL once again positioned the camera in April 2001 for a cloudless pass, this time using the camera's maximum resolution.

Whereas the individual pixels in the 1976 image couldn't show anything smaller than 140 feet across, each pixel in this third image spanned only five feet. NASA's Jim Garvin noted that, in general, you can distinguish objects in a digital image that are about three times bigger than an individual pixel, which meant that everything the size of a small outbuilding on up would be visible. There were nothing but gullies, hillocks, and small outcroppings and bands of exposed rock. Furthermore, this time NASA scientists also directed the onboard MOLA laser altimeter at the feature and compiled a digital, three-dimensional elevation map based on hundreds of measurements with a vertical precision between eight and ten inches. No nose. No mouth. No eyes. Just gullies and hills and rocks.

It is difficult, however, even when comparing the newer images with the old, to disabuse oneself of the notion of a face, so deeply embedded in our brains is the necessity to compile features into bilaterally symmetrical shapes that are recognizable or not, into friend or foe. Between that atavistic templating and our tendency to link discrete visual elements into bounded shapes, we will forever be fooled, at least initially, by first impressions brought to us by remote means. It isn't such a bad survival strategy—assume there's something out there that may be alive, therefore a potential threat, until you know otherwise. But it can also create havoc for the scientific community, much to the delight and profit of Hollywood.

The human cognitive tendency to assemble random noise into patterns that don't exist—a misperception called pareidolia (from *para* meaning false and *eidilon* meaning phantom)—is based on the evolutionary trait that allows us to see the camouflage pattern of a tiger hiding in the tall grass of the savanna, or to recognize the face of a friend in a crowd. It also allows us to construct flying cows in clouds and faces on Mars, and is thus related to our ability to make similes and metaphors. It is boundary recognition carried to a very high level of cognition and it means that the face of Mars will keep changing even after we get there.

When describing the exploration of Mars earlier, I stated that it would take more than photographs to help us understand Mars as a place, that we would have to be there in person and use art to transfer our emotional responses to its terrain to others as part of the process of making it a place. What Elaine and Sam are doing is both part of that process and at the same time an analog activity for that process. (Technically, that makes the filming of her song on Devon a trope, a metaphor that is itself the subject of what it talks about, which is the most complicated kind of metaphor.)

If all we have are aerial images, maps, and photographs of a terrain—which our culture for the most part accepts as scientifically objective documents—we risk not understanding the intrinsic value of a place, but only assessing its utilitarian aspects. Maps are magnificent cultural artifacts that enable us to navigate almost anywhere, but that power tends to make us think that the land it represents is now ours to do with as we please. Unlike metaphors in literature, art, and film, which increase information about a place and thus complicate our view of it, maps and other elevated views simplify things by dropping out information. Cartographic images are reductive and encourage us to believe that the space represented remains nothing more than a specimen, or a commodity available for buying and selling and consuming in a quest for knowledge or resources. No people are visible on a map or satellite photo, and the ground appears to be just an empty room in which to settle an expanding population. It's that kind of thinking that led Canadian politicians to attempt the forced resettlement of Inuit to the islands surrounding Devon. The government assumed that because they had charted the Northwest Passage it was territory that could be owned, and an environment that they understood.

Space art, because it is based on imagination and remotely sensed images, is still utopian by the very fact of its distance from its subject. A much grittier, more realistic, and yet more poetic view of Mars will develop when we're there. The paintings commissioned by Charlie Cockell imagine what it is like to achieve heroic goals of exploration in a sublime landscape. If humans eventually live on Mars, the art they make there will address more subtle realities and drift away from nineteenth-century romantic landscape art. Although still based on human cognitive neurophysiology, it will develop its own styles of representation and abstraction in response to the needs of people on an alien planet. It will be Martian art.

CHAPTER NINE
In the House of Mars

I STAND IN THE door of the cook tent, which is sheltered from the fitfully gathering breeze, and drink the last of my afternoon coffee while Pascal paces back and forth with his notebook in hand, trying to sort out camp logistics for traverses over the next few days. Several of us are hoping to take the Humvee out on a second traverse attempt, but once again the blue sky has given over to mottled clouds, a "mackerel sky" that presages a vigorous and well-organized cold front that's embedded in a low-pressure system and headed straight for us from the west. Snow, in other words.

The weather usually hits Resolute first, and we can track it by listening to the chatter from air traffic control in town. Pascal is killing time before pestering them for an updated report, passing in front of me in his HMP uniform, a pair of tan Carhartt work pants that don't show dust, and a tattered black nylon aviator jacket with pens stuck in the shoulder pocket. Pascal has, I think, always wanted to go to Mars. His apartment is filled with hundreds of books on geology, glaciology, impact craters, astronomy, physics, and Mars—but also Jules Verne novels and model spacecraft. I wonder if it frustrates him that he's ended up working out of what was meant to be the construction site for the habitat up on the hill, versus the three-story lab-and-living complex itself.

Personally, I'd rather be here, not constrained by cycling through a simulated air lock and confined to mock pressure suits. It can be valuable to test exploration and science protocols within the limitations imposed by the exercise, and the FMARS crews conduct useful experiments, but I prefer this kind of fieldwork, where instead of worrying about living in less than a thousand square feet with five or six other people for a month, you're out actually working the terrain.

The sixteen-by-forty-eight-foot tent at my back has been here since the camp was established three years ago and contains a refrigerator, stove, and oven, and a snack table always laden with crackers, dried fruit, cookies, and leftovers from pancakes to pork chops. The propane heater hasn't been fired up yet today, but it will be as the temperature drops. In the meantime, we depend upon the snacks to keep up our metabolism and body heat whenever we're not curled up in a sleeping bag.

In front of me is one of two smaller orange polyvinyl portable shelters, each twelve by twenty-four feet, sturdy tents where workstations are set up to accommodate several people. The second one is to my right and holds all our communication gear and the Hamilton Sundstrand equipment. Walking inside the tents is to enter a distinctly Martian ambience. The light is a muted yellow orange glow punctuated by the bluish screens of laptops; people are swaddled in expedition fleece clothing and down parkas, alternately typing or frowning in concentration, the silence broken mostly by radio communications with Resolute. We might not have the benefit of a hard-shell weatherproof structure in which to live and work—and it is maddening when it rains and condensation on the inside of the tents drips increasingly closer to your laptop—but adding together the enormous dome tent used by the

BEES crew and the greenhouse, plus our personal tents, we have a much larger work space than the hab.

Pascal ducks inside to radio Resolute and to check the updated satellite visible-cloud-cover image that Bill Clancey has pulled up on his computer. He comes back out shaking his head. "It's a no-go for today, and probably for tomorrow," he states. I lift my cup in acknowledgment. If it's off for today, I'll take the chance to spend the rest of the afternoon with the people building the Arthur C. Clarke Greenhouse.

Perhaps the single largest issue facing human explorers of Mars is how to survive there for the five hundred or so days that most mission designs require. Within that issue there are thousands of problems that must be solved, ranging from the constant bombardment of ultraviolet radiation and the ultralethal baths of high-energy particles during solar storms, to the need to protect both planet and explorers from cross contamination. A major logistical crux, however, is the one that caused Franklin to take tinned food on ships: How do you feed everyone over the length of the journey? It's practical to carry at most a year's worth of food aboard a spacecraft, which would cover the six-month trips there and back—but not the eighteen or more months on the surface spent waiting for the planetary opposition to make a return trip possible.

Most Mars mission planners have concluded that a greenhouse on Mars will be necessary, a way of enclosing an artificial environment that also screens out the majority of radiation. Going off-planet for any length of time requires a life-support system with air and water, in essence taking a biosphere with us. In a space suit or spacecraft the environment is created through chemical means, but every plan for even a temporary settlement on Mars has envisioned using biological means provided by a greenhouse. Universities in Canada and the United States are already developing strains of wheat, potatoes, sweet potatoes, and other crops that can be grown in conditions of lower gravity.

The Arctic sun never gets more than thirty-eight degrees above the horizon on Devon, and is constantly in view only from mid-April through mid-August. The brutally short growing season has forced Arctic plants to evolve to take maximum advantage of the sunlight in order to germinate and survive the long winter; food crops from temperate climates, such as tomatoes, take much longer to grow. Among other things, the greenhouse will allow HMP scientists to assess growing food in a season extended by remote telemetry. Just as the Humvee is not meant to be an accurate replica of an actual pressurized rover that would be used on Mars, the greenhouse is likewise an analog test-bed structure—albeit a stationary one—in which a variety of plants, nutrients, equipment, and procedures can be tested.

The greenhouse also represents the much longer-term issue of how humans might get beyond just exploring Mars and actually live there, how we might have Earth on Mars, which is by far the most contentious issue facing the Mars on Earth community. It doesn't help matters that Robert Zubrin espouses settling Mars with industrial terraforming—making another planet more like Earth—under the rubric of manifest destiny, the phrase used by politicians in nineteenth-century America to encourage the imperial conquest of the West by settlers. His choice of words is curious, as the phrase evokes environmental depredations and the genocidal displacement of Native Americans in the name of resource extraction.

When we approach an unfamiliar terrain we seek to construct a visual reality of it that provides scale and directional orientation for ourselves, as well as a host of other sensory clues in space that will enable us to convert it to a place we understand. We also do this with time, assembling various linear narratives, sequences that unfold unidirectionally that we sometimes call fiction and sometimes history, stories into which we can fit our own lives and ambitions, tragedies, and triumphs. Every thought we have about exploring Mars arises from and is invested with stories about life on it, whether life on the scale of the ancient civilizations of Percival Lowell and Edgar Rice Burroughs or life on a smaller scale—the microbes that Charlie Cockell thinks might have existed, or even still exist. Everything we do on Devon Island is predicated on assumptions about one kind of speculative Martian history or another, whether it's practicing how to collect samples without contaminating Martian microbacteria with terrestrial ones, or listening to evangelists proclaiming that we must remake Mars in the image of Earth. All of the histories presume the presence of intelligent life on Mars, past, present, or future.

Speculation that there might be such life in outer space began with Plutarch writing about people on the moon. This idea was adopted by the influential pre-Enlightenment French intellectual, Bernard le Bouvier de Fontenelle (1657–1757), who wrote about aliens visiting the earth from the moon in *Conversations on the Plurality of Worlds*. By the time Nasmyth and Carpenter were finished with their photographs of plaster models, so was Fontenelle's proposition. But Flammarion, embellishing his 1864 thesis that intelligent life must have evolved on other planets, suggested in his 1890 book *Urania* that intelligent life on Mars might consist of deceased humans reincarnated. Alfred Russel Wallace, who along with Darwin proposed the theory of evolution, vehemently disagreed, declaring the planet uninhabitable, but the public sided with Flammarion.

That one of the most famous scientists in the world (who, incidentally, collaborated with Lucien Rudaux) would suggest the presence of intelligent life on the Red Planet opened the door to an entire subgenre of fiction that included H. G. Wells and *The War of the Worlds* (1898), the Edgar Rice Burroughs novels featuring John Carter (eleven in all, starting with *A Princess of Mars* in 1912 and running until 1941), Ray Bradbury's *The Martian Chronicles* (1950), and Kim Stanley Robinson's Mars trilogy (1993–1996).

As we accumulated increasing amounts of visual evidence that there was no past or present alien civilization on Mars, much less a race of superior beings, we shifted the emphasis of our imaginative energies to science fiction based on the future human exploration and colonization of the Red Planet. The idea that Martians might someday actually be a version of ourselves—whether reincarnated or simply born on the planet after it has been terraformed—is where fiction by Bradbury and Robinson lead us. It's a shift from the science romances of the late nineteenth and early twentieth centuries into a harder-edged narrative that parallels the rise of space art. Although I would hesitate to call the "Face on Mars" the last manifestation of the idea that intelligence as we know it has existed previously on Mars—our predilection for finding the "other" being so deeply hardwired into our survival mechanisms—it's now more the exception than the rule.

As part of Bill Clancey's research to understand how robotics can assist human exploration, he has spent seemingly endless hours listening to tapes of astronauts as they conduct EVAs on the moon. He notes that the total sum of our experience in planetary travel is less than two weeks on Earth's moon with a dozen astronauts. Never were they over the horizon from their vehicles, out of touch with Mission Control, or even allowed to stray more than a few seconds from preapproved scripts that dictated their every move, a protocol designed to keep them as far as possible out of harm's way. As he puts it, all our efforts to extrapolate how to explore Mars from those voyages, plus the analog work done on Devon and elsewhere on Earth, are at best science fiction.

The most well-informed and thoroughly realized vision of how we might explore and colonize the planet—in fiction or nonfiction—is the Mars trilogy by Kim Stanley Robinson, which has become a cult classic among NASA scientists as well as Mars enthusiasts. It's a thoughtful and plausible narrative that probes into the social, psychological, and political dimensions, as well as the scientific ones, of both the exploration and settlement of the planet. The dynamic that powers the narrative is the tension between people who wish to terraform Mars and those who wish to leave it as much as possible in its natural state. The "Reds" want to keep people living inside pressure domes and tents, while the "Greens" want to generate an Earth-like atmosphere so people can walk unencumbered on Mars and farm the surface. The adoption of the terms is an ironically distorted reflection of the twenthieth-century terrestrial designations of communists as Reds and environmentalists as Greens, both of them utopians in one guise or another.

The conflict between these competing visions is embodied in large part by a conversation between two protagonists who work together for more than two centuries: Ann Clayborne, a Red planetary geologist, and Saxifrage Russell, a Green theoretical physicist. The mineralogical metaphor of Clayborne's name is obvious; less apparent is the root of Saxifrage, the tiny Alpine flower whose name stems from the Latin word for breaking rocks. Ann and Sax's debate starts while the team is still training in the Dry Valleys of the Antarctic, and continues through a series of revolutions that take place on Mars, political upheavals enabling leaders on Mars and Earth to work out how to balance increasing population pressures and resource extraction issues, while the colonists decide how much of the planet to transform for human needs and desires.

Robinson, a personal friend of many of the HMP people and yet another former member of the board of directors of the Mars Society, confesses that he went back and forth on the issue himself, and that Ann and Sax manifest his internal arguments. The novels were written prior to the formation of the two organizations, but are eerily prescient of the fractures between them. As mentioned earlier, Pascal puts the difference between the Mars Society and HMP as that of church and state, the Mars Society preaching settlement of Mars, the NASA folks discussing exploration.

When I arrived in July 2002, the hab was running simulations and EVAs with Mars Society visitors dressed in their white suits, and a Russian television crew was following them around (which made me wonder how you run a valid simulation in the presence of media). The HMP crew had settled into the tents at the former construction site. The two groups communicated with each other via radio, but in general stayed out of each other's paths. This year we're free to visit the hab if invited. We wave at each other when our ATVs are within visual range, and two of their crew have come over for lunch. We don't share supply flights or offer them routine medical care, but are available during any emergencies, if needed. Now the HMP site is establishing its own solid-walled permanent structure, the greenhouse, which today will receive its first water for the grow tray containing tomato seeds.

I've finished my coffee and Pascal has wandered back into the science tent. The clouds are thickening and the wind is picking up, cutting through my pile jacket. It's with some anticipation that I walk the fifty yards or so to the rounded oblong structure through which moving shadows are visible. The greenhouse is the warmest place in camp, and one of the shadows dimly visible inside belongs to Alain Berinstain, the lead scientist for the project and my interview subject for the morning.

The idea of a greenhouse on Mars dates at least back to an illustration in Clarke's 1951 book *The Exploration of Space,* which included a painting by Leslie Carr of a domed Martian city with rooftop gardens. Willy Ley's *The Exploration of Mars* was illustrated with similar images by Bonestell. The greenhouse crew seeks to extend the growing season from April through October, yet one more variation in a historical process that started when Homo sapiens first cultivated rye in the Near East eleven thousand years ago. As Keith, who is in charge of greenhouse construction, reminds readers in his online article about the greenhouse, humans made the transition from the nomadic life of hunter-gatherers to communal settlers by cultivating plants. The planting of wheat in neolithic Mesopotamia allowed us to form settlements, which eventually provided enough surplus calories in the society to open up time for the pursuit of knowledge, at which point you can claim civilization to have been incubated. We've been extending our range of habitation through agriculture, even before the last ice age ended, by extending our control over the local environment and making possible the settlement of increasingly marginal environments.

An early greenhouse was built by the Romans for Emperor Tiberius during the first century AD and used mica for the windows, glass having not yet been invented. During the seventeenth century, Europeans erected greenhouses to preserve exotic plants brought back from the tropics, and orangeries providing fresh citrus to the wealthy were soon fashionable. When glass became readily available in large quantities during the mid-nineteenth century, the great Victorian age of greenhouses saw the creation of immense public conservatories, such as the palatial edifice at Kew Gardens in London. Glass also made possible the rapid growth in construction of personal greenhouses in America, where it is estimated that today there are now more than three million.

The end point of this technology on Earth is the use of a windowless growing chamber with appropriate artificial lighting inside a modular unit under the dome of the South Pole. Temperatures outside during winter can be minus 100°F—close to Martian in severity—but inside you can recline in a hammock and listen to classical music while all around you lettuce sprouts under grow lights. When I visited the station last year, the tiny facility was producing salads once a week for the entire summer crew of two hundred people.

The HMP greenhouse is a generous twenty-four feet long by twelve feet wide with Lexan panels rising to a gentle peak ten feet above the floor. It's double walled with the literally bulletproof plastic, which is rated to seventy miles per hour, and the structure is secured with aircraft cable attached to anchors going down more than three feet. As I approach the front door, I see that I'm going to have to squeeze through a construction project. Tom Graham and Keegan Boyd, two Canadian graduate students, are constructing an alcove (or "temperature differential dampener," as Keith calls it) around the entrance. Although it's not an air lock, it's reminiscent of one. Keith, who originally trained as a biologist, came up with idea for a greenhouse here in 2001; his partner in www.SpaceRef.com, Marc Boucher, liked the idea, so they sponsored it. Keith ordered an off-the-rack industrial greenhouse made in California, specifying that it be built with a four-foot spread between supports, versus six feet. After test-constructing it in a hangar at Ames, they had its two tons of segments flown to Devon, where they named it after Arthur C. Clarke for his early vision of Mars settlements built under domes.

After I navigate the alcove project and open the door, Alain greets me. The stocky and articulate French Canadian sports a two-week growth of beard, which makes him appear more serious than he is. He has a doctorate in chemistry, specializes in studying the effects of radiation on biological systems, and joined the Canadian Space Agency in 1997. Since then, Alain has managed several missions for science payloads to the U.S. space shuttle and to the Russian space station Mir. Now he's the program scientist for his country's space exploration program, and is their lead planner for Canada's role in future Mars explorations. He's also an adjunct faculty member at Toronto's University of Guelph in the Department of Plant Agriculture, where he researches environmental control systems for greenhouses in extreme environments.

The greenhouse is a translucent structure that looks more like an electronics lab at the moment than a place where anything would grow. The galvanized steel frame is draped with brightly colored wires and cables, all of which feed into computerized control boxes at the far end. The light inside, whether it's sunny or cloudy, is always diffuse. It's not only the warmest place in camp, but on all of Devon Island, and people often take breaks inside from the almost constant pressure and noise of the wind. As a result, there are a couple of folding nylon chairs available, and we settle down for a chat. I ask Alain about how difficult it was to get the project going.

"Funding for the research is relatively easy to get because any new technology we come up with tends to be applicable for the greenhouse industry, which is huge. Here the challenge is the constant swings in weather; dealing with them requires very robust sensors and algorithms. Automation hasn't done that before, and this is probably the most instrumented greenhouse in the world.

"The north side of the greenhouse is insulated from the floor up to five feet with pink foam next to the Lexan, and then a layer of insulated silver foil—the sunlight is weak on that side, so it's a net energy loss there. It's unusual to insulate off one side of a greenhouse from light, but the energy equation here is paramount. The working range of the greenhouse is from roughly 15°F to 75°F. Outside it can range from fifty below to fifty above, so you have to heat and cool it." During summer days, like today, even when it's only intermittently sunny, the two exhaust fans are going almost constantly during the afternoon, sucking out excess heat.

"To balance the temperature, we currently draw power from the diesel generator and use the satellite dish on the hill for telecommunications. But, when the season is over after four or five weeks—which is too short a growing season to get crops—how do we extend that? Last year the objective was to build the structure and monitor the environment inside. This year the objectives are to control that environment and extend the season. For that, we need independent power and communications from May through September, ideally five months, although four would be sufficient.

"You can see that we've erected the first free-standing solar panel outside, and in a few days we'll have a wind generator up, both storing power in batteries. But that means we also have to be very energy efficient. Right now, even with those two fans going, we're running everything on sixty watts—the power of one lightbulb.

"Designing a greenhouse is like doing a spacecraft, and there's a synergy between the engineering. We have to balance what device works when—a heater with a fan, for example—because we have limited power, and that's just like a space vehicle. We're working on heat recovery systems. On warm days, when the batteries are full and the solar panels and wind generator are still generating power, we'll use it to heat a coil in this tub of water and glycol, then run the hot water through a radiator when we need the heat. Next year we'll work on a way to recapture some of the energy from the hot air the fans are blowing outside.

"Over there on the wall behind you is the computer that runs the greenhouse—fans, heaters, the webcam, monitors, and logging devices—and we'll have a small dish, a flat antenna, to send up the data via satellite so we can monitor and control things when we're not here and as it gets darker.

"Behind that insulated wall are several bottles of propane. When half of it is used, we'll stop heating the greenhouse, and that will be the end of the season." He gestures at the two black plastic grow trays next to us, each of which will hold thirty to fifty plants when going. "We'll let things freeze in the one grow tray we have going. In the spring, the sun rises in March after about three months of darkness, and even then it can get up to 113°F in here, though it goes down to 8°F outside at night. But we can start things up with hydroponics in the other grow tray, and by the time we get here in July, have a crop ready to harvest."

In response to my question about how a greenhouse on Mars would differ from this one, he leans back, puts his arms behind his head, stretches, grimaces, then leans forward again.

"For Mars, a greenhouse would have to be pressurized. The pressure there is only one one-hundredth of that on Earth, too low to grow plants. Our group at Guelph can grow plants down to one-tenth of an Earth atmosphere at sea level—but that's still only one-tenth of the problem that Mars has to offer.

"So you have to seal the greenhouse better than this one, which is done just enough to be waterproof. But you have to balance how you pressurize it against the psychological needs of the astronauts, who benefit from time around plants. If you only pressurize up to that one-tenth mark, they would have to wear pressure suits, which would take away all the fun. But, if you fully pressurize, you're using up a lot of resources. Again, it's like a spacecraft.

"Then there's the issue of whether or not the greenhouse should be fully automated; again, there's the benefit of having people working with the plants.

"At this point we're in a proof-of-concept stage, and we're in the engineering phase. Eventually we want to be in a science phase where we open up the facility to people who want to try out different crops and under different growth conditions. We'll be able to partition the facility to do that, a greenhouse within a greenhouse."

After I close up my notebook and depart, leaving Alain and Keegan to resume work on the sensor wiring, I walk back to the kitchen tent and bump into Keith, who's getting ready to walk up to the airstrip and meet a Twin Otter flying in with supplies and more people. It's probably the last flight for a few days, given the weather closing in. Keith pauses to turn and look at the greenhouse in the low light of what is now early evening.

"Look at this," he calls. "Sometimes it's transparent and sometimes it reflects the landscape." Right now it's backlit with sun peeking between the clouds, the structure filled with light, a faceted crystal. Keith is immensely proud of the greenhouse, and will post remote videocam images of it on his Web site even after we've departed.

Because the orbit of Mars is more elliptical than Earth's, the intensity of its sunlight ranges from only as much as 52 percent down to 37 percent of the irradiance the Earth receives at the edge of its atmosphere—a light level about like that inside the greenhouse, actually. The Martian surface, however, has far less moisture in it, hence fewer clouds, and may actually receive more sunlight than prime agricultural land on our own planet. Nonetheless, a greenhouse on Mars will face a number of challenges. In terms of chemistry, Mars has enough carbon dioxide in the atmosphere that it can be mined for the greenhouse, and nitrogen can be derived from human waste, perhaps supplemented with bottled supplies. The lower gravity should be surmountable, but the low temperatures will require heaters, which burn up energy. The biggest obstacle, however, will be the much higher levels of ultraviolet and cosmic radiation, which will require much more efficient shielding than the Lexan provides. Imagine transparent lead.

Besides providing nutrients for human bodies, a greenhouse on Mars will also help stabilize the psychology of the explorers, as Alain pointed out. After the small growth chamber was built inside the Scott-Amundsen Station at the South Pole, the physician on duty during winter-over noticed that people were hanging out there, and he soon started to prescribe time inside for mental health reasons, a practice adopted by the crews at the larger McMurdo Station when they subsequently built their own greenhouse. The cosmonauts aboard Mir likewise reported that they tended to relax when working with their experimental plants. Whatever greenhouse design is adopted for Mars, providing a way for people to use its environment will be critical.

A greenhouse on Mars is either a resistance to or a first step toward the terraforming of Mars, the industrial-scale creation of a self-sustaining ecosystem for the entire planet. If humans go to Mars, they will either be confined by a lack of resources to living inside habitats, or they won't. McKay, Zubrin, and other scientists have authored papers outlining how it would be possible to initiate global modifications to the Martian climate within as short a time period as fifty years with technology we will have this century, if not this decade. Sooner or later every discussion about the future of Mars revolves around the question of terraforming. But, if we decide it is economically feasible and morally acceptable to attempt changing the nature of the other planet, whatever organisms we develop for deployment on Mars will first be grown in a greenhouse.

CHAPTER TEN
The Late Great Church of Mars

THE STORM FRONTS continue from the west for two days before there is a break in the pattern that looks large enough for us to attempt the second journey in the Humvee. The main purpose of the trip will be to take Brian Glass out about ten miles to the west to measure a broad regional magnetic anomaly. Variations in surface magnetism and gravity, which are also found on Mars, give geoscientists such as Brian clues about the underlying structures of the planet. Some of the work can be done as aeromagnetic studies from aircraft—which Brian and Pascal have conducted over the crater since 1999—but Brian is interested in how to conduct the necessary surface studies robotically on Mars, and he's brought with him a portable magnetometer that he can carry in the Humvee as an analog experiment.

Because Brian combines field science with robotics, he understands that Devon provides not just a good geological analog for Martian impact structures, but also for the challenges of how to make things work in a cold, dusty environment where there's no electronic parts store just down the road. When NASA questions why you can't practice Mars in the Mojave, he notes with a grim smile that Devon "enforces a real-world discipline on participants," and how that helps people design instruments that are lightweight and rugged, and also can be repaired in the field. Conducting analog fieldwork in the desert near Los Angeles would be like test-driving the Humvee on a freeway.

This time Jeff Jones will ride inside with them, while I once again act as a scout on an ATV. Jeff is the chief flight surgeon for the International Space Station program at the NASA Johnson Space Center in Houston, and fits to a tee what I've come to recognize as the model of a NASA astronaut: a compactly built individual of medium height who is fit, listens well, and tends to watch everything going on around him while maintaining an air of negligent calm. He has degrees in biology and psychology, is a surgeon who has studied urology and oncology, and does research in space radiation, microbiology, and surgical procedures in microgravity. He's the commanding officer for a Marine air group in the U.S. Navy Reserve, and the flight surgeon for Marine Light Attack Helicopter Squadron 775, Detachment A, medical unit. He also supports the 147th Texas Air National Guard F-16 medical unit, and the 111th Fighter Squadron. In his spare time he backpacks, skis, windsurfs, hikes, and can, I understand, perform reasonably well in most circumstances that involve bouncing balls on playing fields. In NASA parlance: "He has a negative marital status at this time," which comes as no surprise. Who would have time?

Pascal manages to have a NASA flight surgeon at HMP every season, a wise precaution given the occasional injuries that are an accepted risk for people camping in the High Arctic, driving ATVs in rugged terrain, and handling a variety of sharp tools, electronic equipment, and shotguns. Jeff is here to study the hazards and medical issues experienced during exploration, as well as to assess telemedicine tools and protocols for human missions—but like all of the flight surgeons, he also volunteers to take care of any medical emergencies in camp. Just as Brian doesn't have an electronics store down the road, neither does Jeff have an emergency clinic, so it is another analogy for exploration on Mars.

The sky is still spitting a little rain as we leave, but not enough to create any more of a mud hazard than already exists, and we follow the well-worn track up to the airstrip and along its length before heading left and downward, then surmounting the small watershed that separates camp from von Braun Planitia. The track arrows across the ancient playa until climbing the saddle where the *Inuksuk* stands, our signpost to veer left and west on the route used by Pascal when he drove the Humvee in from the coast.

We're headed first for a river crossing a few miles away, which is about halfway to the anomaly. We'll drive in Pascal's earlier tracks as much as possible to minimize the creation of new ruts wherever there's mud to be crossed. Ten miles seems like a long way to travel for a day trip, given our previous record, but we'll be on known ground most of the way, and we've swapped out the triangular Mattracks treads for large overland tires, which means the vehicle can travel much faster. It doesn't, in fact, take us more than a couple of hours to reach the river, even with a couple of stops along the way for Brian to take some readings.

Riding out in front of the Humvee on this trip is nothing like the back-snapping work of the earlier traverse over huge boulder fields. My most important task is to scout the mud patches, driving over them just in front of Pascal so he can judge how treacherous they are. In most cases he can simply follow me, and only occasionally do we have to look for rocky ground nearby to avoid getting stuck. Nor do I have to concentrate as hard on watching for bears, although the usual shotgun is strapped to the front of the ATV. This allows me plenty of time to enjoy the view as the rain lifts and the clouds slowly disperse.

Devon Island is so sterile and forbidding that it takes awhile to come to terms with its austere charms. Even though I grew up in Nevada's Great Basin, the highest and coldest desert of the United States, it takes a few days of traveling in the High Arctic before I recalibrate my expectations of what a healthy landscape should look like. Upon arrival on the island everyone tends to search for signs of life, whether it's a hint of green on a hillside or a vehicle track or a tent. Then those elements become intrusive, and you come to value more the immense and somber stage on which the only other character in the landscape is light.

I find myself moving so quickly along the Humvee tracks that at one point I get too far ahead of Pascal for him to see me. He radios on the walkie-talkie for me to wait, so I shut down the ATV and pull off my helmet. I take a deep breath of silence, then find myself laughing out loud as I remember where I had my first thought of coming to the Arctic. It was while standing in a University of Nevada Las Vegas parking lot on a hot day in May 2001 talking to Frank Schubert, the project manager in charge of the Mars Society's construction projects. Frank is a friendly middle-aged man with curly dark hair, a home builder from Colorado who is a former lead guitarist for the rock band Devo, a member of the Extremophiles band with Sam Burbank, and a Mars enthusiast of the first order.

A group of engineering students was trying to maneuver into place the last of twelve twenty-by-six-foot-six curved sections of the second Mars Society habitat. Each of the sections weighed three hundred pounds, compared to the eight hundred pounds of the fiberglass sections used for the hab on Devon. The foreman in charge of the construction had inexplicably and permanently vacated the site, presumably because there was a discrepancy in how the walls were fitting at the top, but Frank was optimistic they would figure it out.

I was writing a story about a space museum that Paul Fisher, the inventor of the space pen used by astronauts, was planning to build in Las Vegas. Fisher's pen factory is located in Boulder City, just outside Vegas, and while touring through it I met his son, Scott Fisher, who is likewise a space aficionado. The space museum, as it turned out, never got off the ground, but Scott had a new construction company, Built on Integrity, which among other things had developed a proprietary technology for low-cost, energy-efficient housing. It utilized an ultralight steel frame filled with polystyrene foam that was then wrapped in a elastomeric (or self-healing) plastic skin and coated with a ceramic that reflects heat. The segments could be easily broken down and then reassembled by someone working under difficult conditions with only a few simple tools—like an astronaut in a pressure suit on Mars.

The two-story structure had its ground floor of shiny aluminum diamond plate already in place, and the upper floor would be installed once they figured out how to mate the final segments. This particular type of construction wasn't rugged enough for Devon Island, but would do fine for the Utah or Australian deserts, and perhaps even Iceland, all places where the society planned to install additional habitats.

I found it ironic that a prototype for a Mars station would be built in Las Vegas, which already contains the world's largest collection of simulated structures from around the world, such as the Empire State Building, the Eiffel Tower, and Venice's Grand Canal—not to mention a Star Trek attraction. Later I would discover that a developer of interactive tourism attractions had proposed to build a $200 million Mars theme park in nearby Southern California with a simulated Mars exploration base, camp for kids, and a resort and spa for adults complete with a fake crater. The yearning to experience new places provides a never-ending source of tourists looking for analogs of everything from New York City to an alien planet.

The Inuit authorities have several reasons for limiting our presence in the crater. They want to keep what little wildlife there is undisturbed; they require that there be an economic benefit to local communities such as Resolute and Grise Fjord; and they don't want anyone from the outside prospecting for minerals. On top of those quite legitimate concerns, which Pascal does his best to address, is that they have the mistaken idea that Americans want to build a Mars theme park inside the crater. Although they don't have a firm grasp on the fiscal reasons why that would be unlikely on Devon Island, or how repugnant most of us would find such a notion, they aren't that far off in correctly assuming that members of a consumerist society will seek to buy and sell anything, including a place in the hab on Devon Island for a week or two.

In any case, remembering that a hab was constructed in Las Vegas for deployment in a Mars analog setting is an ironic conjunction, but I put the thought aside for the moment when the Humvee churns up beside me. Pascal and Jeff examine the map, then declare that the river is just over the next ridge. A few minutes later and we're parked on a bluff above the watercourse, sunlight glinting off rills and wet stones. It looks like there's a ford shallow enough for us to make it across to a gravel bar in the middle. We can drive to the downstream end where there's another crossing, thus avoiding pools too deep for the ATV.

As we're sussing it out, the radio crackles to life with Steve Braham's distinct voice. When Pascal is out of the camp, Steve is next in command, and it's obvious even from outside the cab of the vehicle that stress is pitching his voice higher than normal. It turns out that one of the Inuit students has had what appears to be a seizure of some kind, and he'd like Jeff to return to camp to examine her. There doesn't appear to be any immediate crisis—she's resting in a tent and says that she's had these before, that she'd forgotten to take her medication, which they've now given her—but everyone is worried about her well-being. Jeff gets on the radio for a lengthy consultation with Steve and other camp members. There's nothing I can add to the conversation, so I ride down to the riverbank to take a closer look at the crossing spot.

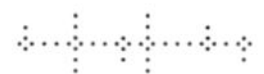

It's good to sit by running water, a meditative tradition in so many cultures worldwide that I wonder what the mechanism is in our neurophysiology that produces a sense of reverie in response to white noise. Cognitive scientists who have measured brain activity in serious practitioners of meditation, including Buddhist monks and Catholic nuns, state that when they fall into their deepest sense of being one with the world, it appears as if the brain is shutting out much of its external input and running some kind of internalized feedback loop. White noise has the effect of canceling out other aural input, so perhaps the response is related. I don't know.

I do, however, welcome a chance to contemplate an issue triggered by thinking about the creation of themed environments in Las Vegas: the terraforming of Mars. Las Vegas is an extraordinary presumption on Earth. Located in the northeastern corner of the Mojave Desert, a region that receives an average of only five inches of rain per year, it has nonetheless remained one of the fastest growing cities in America for two decades, and its casino fountains on the Strip are prodigious and legendary. Its nearly two million citizens use more water per capita than any other urban residents in America, most of which is for inappropriate landscaping, such as green lawns. I'm tempted to call Las Vegas the most extreme example of a landscape makeover in the history of the world, yet it is nothing compared to what some people propose doing to the deserts of Mars.

The first person to propose terraforming may have been the classic science fiction writer Olaf Stapleton, who in his 1930 epic *Last and First Men* proposed applying electrolysis to the supposed seas of Venus to generate oxygen. Jack Williamson, however, is given credit for first using the term in *Seetee Ship* in 1942 (written under the pseudonym of Will Stewart). The first novel to terraform the Red Planet was *The Sands of Mars* by Arthur C. Clarke in 1951, and by 1973 Carl Sagan was suggesting that we could melt its polar ice caps by covering them with either dark dust or plants.

Chris McKay at Ames wrote a science paper three years later, "On the Habitability of Mars: An Approach to Planetary Ecosynthesis," that offered the introduction of microbes to produce a living ecosystem, a project he estimated would take something on the order of one hundred thousand years. His use of the word "ecosynthesis," coined by a colleague at Ames, Bob McElroy, was significant in that McKay believed it might be possible to restore a carbon-dioxide atmosphere on Mars that would be sufficient to establish (or reestablish) microbial life, a state of self-sustaining "ecopoiesis." He did not think it feasible to transform Mars into an Earth-like environment, which would require holding a much denser atmosphere in place.

During the last decade of the twentieth century a vigorous debate over the issue was in full swing. While the science writer Martyn Fogg was penning his technical textbook on methodology, *Terraforming: Engineering Planetary Environments,* Kim Stanley Robinson was finishing up the Mars trilogy. The idea of terraforming in the public imagination had two sides: It was the kind of science we would need in order to live on other planets—but was exporting ourselves and the mess we'd made on Earth a good idea? How would it be possible to avoid screwing up the environment of another planet where we ourselves had not evolved?

Robert Zubrin grew up watching the Apollo missions on television and building his own model rockets, and after college became a high school science teacher. During the 1980s he went back to graduate school to earn degrees in aeronautics and nuclear engineering and found a job working on thermonuclear fusion, which he considered the technology of the future. When funding dried up, he found a job at Martin Marietta Astronautics in Denver, the company that had contracted to build and run the Viking missions. Zubrin soon became disillusioned with the expense and complexity of NASA's subsequent 1990 proposal for sending humans to Mars—a $450 billion plan requiring enormous stations in space and on the moon—and he participated in a team charged with developing an alternate scenario that came to be called "Mars Direct."

The NASA and HMP scientists with whom I've talked say that the plan, although it contains some glaring omissions and flaws, has simplicity and boldness in its favor, and they praise it for improving NASA's ability to sell a human Mars mission to Congress. First, an unmanned "Earth Return Vehicle" (ERV) is sent to Mars with a tank of hydrogen, a chemical plant, and a modest nuclear reactor. Once there, it mines carbon dioxide from the atmosphere and mixes it with the hydrogen to produce both oxygen and fuel for the return trip. Twelve months later a crew of four is launched to Mars, where they spend a year exploring the planet by rover. They return using the tiny ERV. Cost: $7 billion. Dan Goldin, the "cheaper, faster, better" administrator, liked the idea, and after three years of stiff internal debate NASA adopted an expanded version as a design reference mission from which to do further planning. The compromise increased the crew to six members, made the rover larger, and increased the size of the ERV so people wouldn't kill each other in transit. Cost: $55 billion.

NASA engineers realized that even with the amplified budget there were some serious shortcomings still in the plan. For example, it shrugged off the effects of long-term exposure to lessened gravity as only temporary, and to radiation as causing only an additional 0.5 to 1 percent increase in the 20 percent chance each person has, on average, of developing a fatal cancer on Earth. Based on the experience of International Space Station and Mir crew members, who take up to six months to recover from their tours of duty in space, some doctors estimate that it could take as long as a decade to recover from the much longer Mars mission. And radiation remains a potential showstopper for a human mission to Mars.

According to the NASA scientists at Brookhaven National Laboratory, an average American absorbs about 350 millirems of radiation a year, received from sources as diverse as cosmic rays, radioactive rocks, and medical x-rays. The *Apollo 14* astronauts took in 1,140 millirems—three years' worth of exposure—during nine days on the moon. *Skylab 4* astronauts got hit with 17,800 during their eighty-seven days in orbit around Earth, right at the limit where clinically measurable symptoms start to occur. A round-trip to Mars would blast astronauts with 130,000 millirems during the mission, the equivalent to four hundred years of exposure on Earth. And unlike the relatively benign gamma rays we mostly get on Earth, this would be from ions all across the lower half of the periodic table traveling almost at the speed of light with much heavier charges.

The likeliest shield while in transit might be the water that the mission would have to bring with it in any case. The astronauts' living quarters would be surrounded by a cylindrical water tank. On the planet they would have to live underground and have a way, while out on traverses, to throw up shelter from unexpected solar storms, which could cook them in minutes. But Zubrin is an engineer, not a scientist, and I have often found there to be a key difference between the mentalities. A scientist will remain skeptical until proven otherwise; an engineer will assume anything can and will be built. "Go fever" of a sort.

Zubrin has always insisted that the private sector could mount a mission to Mars more efficiently than the government, and in 1996 he left what had become Lockheed Martin to form his own company, Pioneer Astronautics. That year he also published *The Case for Mars: The Plan to Settle the Red Planet and Why We Must,* the first book of several he would write to promote his ideas. The next year, using addresses from fan letters, he started to gather support to form the Mars Society, enlisting the help of everyone from Chris McKay to Kim Stanley Robinson. In August 1998 the first conference of the society was held, and Pascal was there to give a report from Devon Island.

Zubrin underpins his case for smaller and bolder efforts by using historical analogy. He notes that Franklin's large and expensive government expedition attempted to force its way through the Northwest Passage and failed, but that other explorers on smaller expeditions used Inuit dogsleds to make the journey on foot, and that Amundsen finally sailed the passage with a small boat and a minimum of personnel—the Arctic Direct, if you will. It's a nice analogy, but it's also hopelessly inaccurate to cast matters as, forgive the pun, polar opposites.

In 1999 a University of Washington economics professor, Jonathan Karpoff, published a paper that analyzed thirty-five government and fifty-six private expeditions attempting to locate and sail through the Northwest Passage between 1818 and 1909. The government expeditions lost just over half their ships, and even the privately financed ones lost a quarter of them. Karpoff noted that the private ventures were more highly motivated than the lavishly financed public efforts, and that they explored more new territory, introduced more new technology, and lost fewer lives. He concluded that the government expeditions were less efficient for three reasons: The leaders of the public expeditions were "relatively unmotivated and unprepared for Arctic exploration; leadership initiation was separated from implementation; and the leaders failed to adopt new information relevant to Arctic exploration, from clothing and diet to organizational principles." Karpoff went on to suggest that NASA should sponsor large prizes for private enterprise to attempt the exploration of space.

On the face of it, the study would seem to support Zubrin's analogy. I don't doubt the professor's analysis, and I certainly applaud NASA's recent openness to follow the lead of Microsoft cofounder Paul Allen and others to develop private space travel. But I would submit that two factors should be taken into account before blindly applying this analysis to the exploration of Mars. One is the historical synergy between public and private expeditions; the other is the learning curve.

It is true that large government polar expedition programs tended to be more conservative in their efforts than forays made by others, but by their very ability to absorb failures, they paid for lessons that were then adopted by the private parties. It was a dynamic relationship, and no single method of exploration ever opened up significant portions of the planet's surface. It has always taken a mix of government efforts fueled primarily by strategic motives, and then secondarily by scientific ones. The expansive government voyages to discover the Northwest Passage included those led by Captain James Cook, John Ross, and his nephew James; private enterprise contributed through guides working for the Hudson's Bay Company, and the intrepid whalers sailing both north and south; only then did independent explorers such as Amundsen make their mark.

Furthermore, the relationship between past and present expedition planning and implementation continues to be a dynamic one. NASA learns from the past and its own failures as time goes on, hence its ability to adopt some of Zubrin's suggestions. Furthermore, I doubt anyone would think that astronauts commanding missions today are equivalent in any way to the sometimes uninspired and unprepared naval officers of almost two centuries ago. See Jeff Jones's résumé, page 184. The very presence of someone such as Bill Clancey—a professional ethnographer who works out of Ames at both the HMP and the Mars Society habitats here and in Utah—is an example of how self-aware NASA is about learning from exploration simulacra even as it conducts them.

In any case, by the time the first Mars Society conference had concluded, its members had hammered out a reasonable mission statement for the society: "Its purpose is to further the goal of the exploration and settlement of the Red Planet. This will be done by broad public outreach to instill the vision of pioneering Mars, support of ever more aggressive government funded Mars exploration programs around the world, and conducting Mars exploration on a private basis. Starting small, with hitchhiker payloads on government funded missions, the society intends to use the credibility that such activity will engender to mobilize larger resources that will enable stand-alone private robotic missions and ultimately human exploration."

The Mars Society made a valuable contribution both to NASA's thinking and to public advocacy for Mars exploration, despite the somewhat one-sided nature of Zubrin's historical analogy. When Zubrin goes beyond exploration to discuss colonization, however, he is blithely ignoring the very lessons of history to which he has been arguing for us to attend. He touts Mars as a planet with "spectacular mountains three times as tall as Mount Everest, canyons three times as deep and five times as long as the Grand Canyon, vast ice fields, and thousands of kilometers of mysterious dry riverbeds," and describes it as having once been a warm and wet environment with water still readily available in the permafrost just beneath the surface. His language, while mostly accurate, is also reminiscent of the pitch used by the boosters selling western expansionism during the nineteenth century, men such as U.S. Senator Hart Benton (the patron and father-in-law of the explorer John C. Frémont) and William Gilpin (the first territorial governor of Colorado and a rapacious landowner). These were men who adopted the rationale of manifest destiny—what they considered to be the obvious and self-evident necessity to link the east and west coasts of America so that the flow of capital could proceed unimpeded around the globe. They touted the climactic scenery of Yellowstone, the Rocky Mountains, and the Grand Canyon to potential emigrants, while downplaying the presence of any potentially fatal desert in the region. They even espoused the pseudoscience that claimed rain would literally follow if you plowed the land.

"A new branch of civilization" is what Zubrin thinks Mars will give us, a refuge while various small nations on Earth squabble over diminishing resources. Writing in his oft-reprinted polemic "Mars: The Frontier Humanity Needs," he postulates: "Western humanist civilization as we know and value it today was born in expansion, grew in expansion, and can only exist in a dynamic expanding state. While some form of human society might persist in a nonexpanding world, that society will not foster freedom, creativity, individuality, or progress."

Not only does he attempt to claim that the only viable society on Earth is essentially that of the United States, but never does Zubrin question whether it is right to colonize Mars. He considers terraforming technologically feasible given the application of industrial nuclear power; it is only necessary to have the capital to do so, and for that he proposes to mine the planet for the exports Earth will crave. He again follows the nineteenth-century American model of the frontier: that it is a wasteland into which excess population can escape and from which resources should be extracted. As a corollary that I find painfully comical, he contends that real estate speculation on the Red Planet is, at this stage, not unreasonable.

Here's Zubrin on making Mars fully habitable for humans, a process that he thinks would take a tenth the amount of time hypothesized by McKay for ecosynthesis alone: "Some people consider the idea of terraforming Mars heretical—humanity playing God. Yet others would see in such an accomplishment the most profound vindication of the divine nature of the human spirit, exercised in its highest form to bring a dead world to life. . . . Indeed, I would go farther. . . . Countless beings have lived and died to transform the Earth into a place that could create and allow human existence. Now it's our turn to do our part . . . if we can terraform Mars, it will show that the worlds of the heavens themselves are *subject* [his emphasis] to the human intelligent will."

Needless to say, when Lowell Wood, a nuclear weapons scientist from Lawrence Livermore National Laboratory, used the term "manifest destiny" to justify terraforming during a conference session on the subject, not a few in the audience became upset. Zubrin quickly adopted the phrase. Both Wood and Zubrin espoused the viewpoint that it is correct to value humans above nature, and that bringing life to a lifeless planet is a moral obligation. More than a few people would find themselves drifting quietly away from the organization in the months and years ahead, dismayed by a rhetoric that was naive at best. Zubrin characterized the disagreement as merely one over "terminology," apparently oblivious to the lessons of history. Unlike the characters in Kim Stanley Robinson's novels, which he professes to admire, Zubrin merely wants to extend an old paradigm for exploration, rather than invent a new one.

The evocation of the divine in the language of nineteenth-century politicians is to be expected. To find it in the work of a nuclear engineer at the beginning of the twenty-first century confounds me. Writing in his book *Mars on Earth* about the Mars Society's Flashline Arctic Station, Zubrin first mischaracterizes the definition of the word terraforming as "planetary engineering with the specific intent to transform a sterile world into a dead one." He gives himself away when he clarifies the "core idea" behind it as going "back to the Judeo-Christian notion that since man is the image of God, man should attempt to continue God's work. Terraforming is continuing the work of creation. So whether aware of it or not, would-be terraformers engage in a design project to build a planetary cathedral celebrating the divine nature of human reason."

Such a position—I hesitate to call it reasoning—acknowledges only Western traditions as a valid basis for civilization and reveals a narrowness of understanding that is at odds with even the concept of broadening the reach of humankind to another planet. The rhetoric also evokes a Eurocentric rationale that has been given for every kind of slaughter imaginable.

The phrase "manifest destiny" was developed during the nineteenth century by New York journalist and editor John O'Sullivan. In 1839 he wrote an article in his magazine *The United States Democratic Review* titled "The Great Nation of Futurity," which contains both the seed of the phrase and language akin to Zubrin's:

> The far-reaching, the boundless future will be the era of American greatness. In its magnificent domain of space and time, the nation of many nations is destined to manifest to mankind the excellence of divine principles; to establish on earth the noblest temple ever dedicated to the worship of the Most High—the Sacred and The True. Its floor shall be a hemisphere—its roof

> the firmament of the star-studded heavens, and its congregation a Union of many Republics, comprising hundreds of happy millions, calling, owning no man master, but governed by God's natural and moral law.

In 1845 O'Sullivan penned an unsigned editorial that appeared in the *New York Morning News* that tied his rationale directly to the tradition that Zubrin is alluding to from the biblical Genesis: "It was the nation's manifest destiny to overspread and to possess the whole of the continent which Providence has given us for the development of the great experiment of liberty and federated self-government entrusted to us." Zubrin sees Mars as literally being populated by a new generation of American frontiersmen.

It's not that I'm against exploring Mars; I'm all for it, and agree that it is a healthy goal for the human race (not simply Americans). And I can be persuaded that there might be some conditions under which it would be desirable to attempt some version of ecosynthesis. But I can't find any reason to assume that it is our unquestioned right to undertake such endeavors, much less one granted us by a hypothetical divinity. In fact, I can find every reason why we should avoid such logic, given the lessons of nineteenth-century American expansionism.

My peregrinations by the river are interrupted by a squawk from my walkie-talkie; it's time to return uphill to the Humvee. Once there it's obvious from the glum expressions that we're going back to base. Although Jeff was able to identify clearly what the problem is—a relatively common and minor ailment that should pose no problem for our student intern now that she's back on her meds—after an hour or so of debate it's become obvious that for everyone's peace of mind (including the patient's) we need to return. It's a fine lesson, I think, in how the decision-making process in exploration sometimes has to answer to emotional needs as well as logistical and scientific ones.

Half an hour later, as we're driving back over the ridge and onto von Braun Planitia, we pass three figures in white mock pressure suits examining the *Inuksuk*. It's a moment pregnant with meaning and association, the crew members from the Mars Society hab clustered around the monument like one of those scenes from *2001: A Space Odyssey,* where at first it is the apes and then the astronauts who similarly surround an alien obelisk in awe and worship. Here the field is reversed and turned inside out, the mock astronauts taking pictures of a pretechnological artifact erected by their contemporaries.

I stop my ATV for a moment to look at the figures, the rock in front of them glowing orange in the low sun, and realize that the *Inuksuk* is not just an artifact that signals a physical orientation in space, but also a place marker that helps us navigate in the stream of events, in history. First, as a class of Inuit object, it symbolizes the presence of people in the land, the necessity to construct a stand-in for another human in a land where it can be supremely difficult to know one's position. This particular *Inuksuk* also represents the lost life of an astronaut, and thus stands for the difficulties inherent in space travel. And now, this conjunction of hab crew members with the life-size monument draws a metaphorical line from the history of the Inuit in the Arctic to that of humans on Mars. It's an arc of adversity—the inability of humans to "subdue" this fierce land into landscape for any significant length of time; the impossibility of Western technology to "conquer" space; the potential failure of people to learn from what is more than "mere" terminology throughout history.

CHAPTER ELEVEN
The View from Olympus Mons

THE MORNING I'm due to fly out of Devon, I strike my tent and pack it along with spare clothes, sleeping bag, and foam pad into a huge yellow duffel bag. I heave it onto a pile of gear we'll shift over to the airstrip with an ATV, then shoulder my rucksack, heavy with laptop, notebooks, and still more clothes, and walk over to the kitchen tent. I stow my pack behind the stove and let Joe know that I'm just going up to the crater rim and I don't need a shotgun. Back outside I stuff hands in pockets and trudge over the small ridge behind camp. On the other side I cross the diminutive Lowell Canal. A small generator pumps water steadily out of the stream and sends it back to camp through a hose, the noise annoying but always reassuring.

Ten minutes of steady plowing up the loose talus of Haynes Ridge puts me on the rim of the crater. A bank of clouds hovers over the north shore of the island where the Haughton River debouches into Bear Bay and Jones Sound. Invisible to me from here, Ellesmere Island sits across the narrow strait. On the far edge of that much larger island is a rock, newly exposed by the melting ice, that is the northernmost point of land in North America. From there nothing stands before you but the great revolving gyre of Arctic ice and the North Pole.

It's only July 29, but already the two weeks of good weather that the camp receives each year have come and gone. Cloud coverage and wind speeds will start to increase from week to week, and the days grow progressively shorter until everything is frozen once more. There's a short window of good weather for a flight around noon, and most of us will fly out. Pascal, Joe, Steve, and a handful of other people will stay for another few days to take down the large camp tents, which along with the ATVs will be packed inside the kitchen. Once everyone is gone only the autonomous greenhouse will continue as a warm structure into the fall, the tomato plants germinating, monitored by Keegan and Alain from down south.

A few hundred feet below me the sloping interior of the crater is wet with meltwater, all headed downward to the bottom of the crater and the river. I can see across the silvery breccia of the depression to the far rim, its steep cliffs and canyons hinted at by darker folds in the land. Last summer I'd had a late traverse across the crater to that side. Although I've worked in many of the places on Earth that have served as Mars analogs—the Dry Valleys, the back side of the Himalaya near Tibet, the Sedan Crater on the Nevada Test Site, the hot lava beds of the Kilauea caldera in Hawaii, and the cold cinders of Ubehebe Crater in Death Valley—and although all those environments have elements in common with Martian terrain, most especially the Dry Valleys, none of those journeys had the poignancy of the Haughton traverse.

No one had been inside the Inuit Owned Land of the crater since 1999, the last year of the original HMP camp, and Oz needed at least one more foray to complete the fieldwork for his dissertation. Pascal had been lobbying hard since the spring to get him permission, and when it finally came it was almost the end of July. This was the traverse that I had crossed my fingers to make someday when visiting Charlie's field camp. Our team consisted of Alain, who went as the official Canadian government representative, Oz, myself, and the requisite Inuk, in this case our young friend Jeffrey, the *Star Wars* devotee. While Oz and Jeffrey rode on ATVs, Alain and I drove Priscilla, the project's two-person, six-wheeled amphibian vehicle called an Argo. A plastic bathtub with wheels, it proved a bumpy ride, but the six wheels had a tenacious grasp of almost any terrain.

We headed out in the morning under cloudy skies in a light rain, the crater somber in a palette of dull grays as we breached the rim. The vehicles rolled under the Breccia Hills where Charlie had worked with the Japanese film crew, then by his little field camp and down to Rhinoceros Creek. Instead of turning right, for the first time we were allowed to head left toward where the creek joined the Haughton River. Alain and I stopped midway down the valley while Oz and Jeffrey went on ahead to do some serious geologizing. With rock hammers in hand, Alain's task was to measure some of the local topography with a more recent addition to the geology toolbox, a lidar, or light detection and ranging instrument, which is closely related to the MOLA laser altimeter aboard the *Mars Global Surveyor*. Alain carefully unpacked the off-the-shelf surveying unit from its insulated plastic crate, the instrument a nondescript white square box that he attached to a tripod.

Alain's lidar measured the minute differences in the time it takes light from individual pulses to scatter back after it hits the target. Sending out two thousand pulses per second spaced about a half-inch apart, a single scan of the hillside would take forty-five minutes, creating a virtual cloud of data points that Alain would then translate on his laptop into an accurate three-dimensional map of its surface. The algorithms for the translation would sort the data points into polygons to create a jointing interface, a nice metaphorical parallel to the literal polygonal sorting of the ground under our feet caused by frost heaving.

It was still drizzling as Alain calibrated his unit to the steep breccia face on the other side of the creek, a hillside around three hundred feet away. A lidar can map the movements of glacial ice, serve as a precise altimeter, and perform any number of miraculous cartographic feats, including help a lander achieve a precision landing in difficult terrain. It cannot, however, see through rain, the droplets of which could block the light scattered back. But the precipitation let up, Alain hit the start button, and he commenced to walk slowly in circles while looking at the ground. I took off toward the ridge behind us, plodding steadily uphill over the patterned ground.

Polygons, ranks of solifluction lines, circular frost boils—all the surface indicators for a periglacial environment were there, as they are on parts of Mars. Small patches of moss every few hundred yards marked where an animal—usually a lemming—had died and fertilized the ground, a distinctly un-Martian phenomenon, at least as far as we know. The evidence for water on Mars is that it's all underground, and if there's life on the Red Planet, that's where it's most likely to be found, safe from the radiation, the sterilizing peroxides, and temperature extremes of the surface.

From atop the ridge I could see back toward Haynes Ridge and the hab, then to the opposite side of the crater on the far side of the Haughton River, which was itself hidden from my vantage point. As the clouds began to break, the breccia lightened from gray to silver. The hills and impact rings seemed to ripple in the light, almost as if the land itself were waking. The view was at once dendritic, fractal, chaotic, yet deeply patterned by geological processes. You could read it, but it was like interpreting the lidar data. You had to apply the algorithms of experience to interpret the visual input in order to even see the patterns, much less understand them.

I sat on the damp hillside and thought about Zubrin's proposal to use mining operations to finance human habitation on Mars, and compared it with a distinctly opposing viewpoint, that authored by Stephen J. Pyne, one of America's preeminent historians of exploration. Pyne, who spent months in the Antarctic writing a natural and cultural history of the continent, trained under William Goetzmann, the dean of Western exploration history. One of Goetzmann's contributions to the field is his definition of eras within exploration. He defined the "First Age" as that of the great "voyages of discovery" starting in the fourteenth century, circumnavigations of the globe that culminated with James Cook in the late seventeenth century. The Second Age, he ruled, was that of the continental traverses across Africa, South America, and the American West.

Pyne is thus more than conversant with the nineteenth-century philosophy of manifest destiny, and hypothesized his own Third Age of exploration, which he defined in his Antarctic book, *The Ice,* as one that after World War II began to rely increasingly upon remote sensing, such as the lidar that Alain was deploying below me. When asked to write an essay for the Smithsonian book *Space: Discovery and Exploration,* he acknowledged that American national identity is tied to geographical discovery:

> For some time now, the dominant creation myth of America has followed just this formula, that it has resulted from the

> encounter of Old World civilization with New World wilderness and that from this sustained "discovery" derives America's greater purpose and moral vitality. . . . Without continuing discovery the West would presumably suffer a crisis of courage, succumb to spiritual malaise, perhaps even wither into decadence and timidity. . . . The voyage of discovery became a vehicle for progress; the future belonged to those nations with the means and fortitude to explore.

He then reminded readers that pioneering hadn't just brought back new knowledge, but inevitably had destroyed that which was being explored, whether it was through the disruption of ecosystems through resource extraction, the wasting of native peoples by the importation of disease, or just outright conquest. Pyne believes that exploration is an essential, even a defining cultural artifact of Western civilization, but that the nature of the contemporary endeavor should not attempt to mimic the voyages of Cook around the Pacific or the expeditions of Alexander von Humboldt across South America. He proposed that Third Age voyages be done by proxy.

> In past ages discovery had to be done by people. There was no other option by which to learn the language, to record data and impressions, to meet other societies and translate their accumulated wisdom. . . . But humans do not have to be physically present in the Third Age, and there are good reasons for arguing they should limit their voyaging to the near-space of Earth. Robots, remote sensing, and television can get to the critical environments, record the sights, and take the necessary measurements, and do these jobs for far less cost than required by manned voyages.

Pyne noted that, because humans would be confined to spacecraft, rovers, and pressure suits while exploring other planets, only vision would be left as a perceptual mechanism; and he maintained that television would ably serve the purpose while bringing back "awe, information, excitement, and perhaps wealth without exporting death, contaminated biotas, and moral qualms."

If I found Zubrin's plans disturbingly evangelistic, then I found Pyne's approach too theoretical. Six months before coming to Devon I had spent several days in the most remote of all the Dry Valleys, a place where the oldest ice in the world is said to exist, a moraine-covered glacier that may be as old as eight million years. A Twin Otter carrying a lidar unit mapped the surface of the glacier within a fraction of an inch, creating a baseline to measure its infinitesimal movements. But on the ground we had to use shovels, trowels, and finally a chainsaw in order to get deep enough in the ice to find the ancient microorganisms present, a job that robots would have been unable to accomplish. Remote sensing complemented, but could not supplant, the personal fieldwork. The same was true in Haughton Crater. Alain was modeling the terrain with a remote sensing device, but it took human beings, Oz and Jeffrey, to climb the slopes together and gather the samples, an action that a robot wouldn't be able to perform. All day long as I bounced along in the Argo, I weighed the relative merits of the two extremes—Pyne's vision of teleoperated or semiautonomous machines driving slowly around Mars versus Zubrin's converting it into the fields of corporate agriculture.

When Alain was finished making his second scan we piled everything in Priscilla and proceeded to follow the creek downstream. When the valley pinched to a canyon, Alain scared the hell out of me by plunging down a steep gravel embankment and directly into a deep azurine pool of Arctic water. Priscilla calmly motored down the stream, across the pool, and up the other bank. Alain laughed at my expression, a compound of open-mouthed surprise and wide-eyed awe. Driving on water was a new experience for me.

Shortly afterward we found Oz and Jeffrey, who were happily covered in mud, having successfully excavated a number of samples. Together we proceeded down to the confluence of the stream and river, turned right, and soon stopped on a beach of water-worn stones across from cliffs undercut by pools of turquoise water. A small waterfall shot off the lip of limestone near a shallow grotto. Oz was interested in examining the cliff, so he hopped into Priscilla with Alain and they drove sedately across the river to the overhanging rock. While we waited ashore, Jeffrey practiced his Jedi warrior routine along the gravel. I sat on the riverbank examining pieces of shocked gneiss. It still amazed me how the minerals in the rocks had been vaporized away by the impact of the body that created the crater. When the Argo returned, Oz reported that they'd spotted a two-foot-long fossil, "some kind of ammonite-related animal, probably Ordovician, preimpact. It was sitting up too high for us to get at it, though."

"Ordovician?" I asked.

"Yeah, four hundred million years give or take." I blinked. The last glacial maximum ended here between ten and eight thousand years ago, leaving behind a remnant of the Laurentide/Innuitian ice sheet as a small ice cap covering the eastern third of the island. That event is already deep enough in the past that it's at the very edge of our imagination. The crater was thirty-eight million years old, the fossil roughly seventeen times that age. The shocked gneiss that I held in my hand, a piece of the ancient basement rock of North America, was 2.5 billion years old—more than a hundred times older than the crater.

And life on Mars? Life on Earth arose more than 3.5 billion years before present, when conditions were probably similar on the two planets, a preoxygen environment prior to microbes evolving into photosynthetic organisms. Figuring out whether or not genetic material was exchanged between the two planets, or if it arose on Mars independently of that on Earth, was the key question. Fossil nanobacteria billions of years old that may exist on Mars—where they may not have been eliminated by tectonic action, as on Earth—could be as old as the rock I held in my hand. Finding them would answer the question.

Once back ashore, Alain and Oz interrupted my peregrinations by motioning us upriver toward the original HMP camp. Jeffrey and I followed on the ATVs, the canyon widening back out into a valley. About a half mile of driving on flat terraces brought us to the edge of a broad bench where the camp had been pitched, a raised area of several acres with a fine view of the river in both directions. By then most of the clouds had disappeared in the late afternoon light, only a few shadows chasing around at the far southeastern end of the crater. Several rock rings stood above the gravel, anchors used to peg down the tents that had been there several years previously. I bent down to examine one of the rocks, then another and another, amazed to find that they were the largest shatter cones I'd ever seen, the impact inscribed into their sides like the radiating lines of multiple sunbursts. Stromatolites—the fossil remains of bacterial mats from the lake formed after the impact—and ancient corals were also represented, as were enormous planar crystals of milky white selenite. Not far away was an outcropping of the transparent gypsum, the upper end of a natural hydrothermal pipe through which the mineral had precipitated.

The site was an open warehouse of wonders, and I could see why Pascal, when he had sent me off in the morning, had been smiling so broadly. He has remained very fond of the place for obvious reasons. Good science, great views, water and clear landing sites nearby. The camp on the rim of the crater gave more ready access to other kinds of terrain, but clearly the deep interior of the crater was extraordinary. Finding the old campsite was like visiting an abandoned exploration site on another planet.

We took a roundabout way back to camp that day, climbing over a notoriously muddy and slippery pass in order to drive by the edge of the Lowell Oasis that I'd spied some days earlier from above Charlie's camp. At first I thought what I was seeing was a trick of the light glinting off standing water created by a braiding stream—but it was, in fact, tundra, the only patch in the center of the island, a growth made possible because lake sediments provided a rich enough ground for the colonization by hardy Arctic plants, and eventually a few animals, including the small herd of musk ox once sighted here. There was no evidence of the animals that day, but it still astonished me to see grass in the low point of the crater, and it did not at all diminish my sense of how wild and desolate the rest of the island was. In fact, by virtue of contrast it only increased the marvel of it.

What made the traverse so meaningful, so laden with a sense of wonder and loss, was the existence and fragility of the oasis at the bottom of a catastrophic landscape. It wasn't just the contrast with the incessant sterility on the rest of Devon Island, but the vivid proof that the kind of cataclysmic event that occurs on all planets would actually make a terrain more fit for life, that it had created a bowl in which evolution in the Arctic would proceed in a slightly different direction than elsewhere. The evidence flooded in as the color green, the smell of grass, the sound of running water, an increase in warmth and humidity. The multiplicity of sensory responses was almost too much to take after relying so much on just visual stimuli during the field season.

Now a year later as I look down from the top of Haynes Ridge before flying out, I can remember where the river runs by identifying the ridges we drove that day. I can trace a route that traverses what is almost, save the oasis, nothing but pure geology. Chris McKay, when talking about ecosynthesis, reminds us of the distinction between nature and life: On Earth, we consider them synonymous, but on Mars we might find only rock—nature but not life. That's the poignancy that I had experienced that day, the difference between the two, and how valuable each was on its own and with each other.

If that is a lesson to be learned on Devon Island and in other analog environments such as the Antarctic, what does it imply about life on Mars and terraforming? It's taken me a year to mull over options to the positions of Zubrin and Pyne, and it was not of much surprise to find that it was Kim Stanley Robinson and one of his colleagues, the respected author Greg Bear, who had come up with two compelling ideas.

If colonization defines one end of the spectrum in Mars exploration, and robotic exploration the other, then Robinson has framed a middle way that is developed by following a decision tree. Send people to Mars, but don't do anything hasty. Keep our planet's genome from contaminating Mars until we know if there is life there now, or if it has been there in the past. If there is indigenous life, microbial or otherwise, we should try to kill any terrestrial bacteria we might have brought with us. If Martian life is fossilized or in suspension, we might consider bringing it back. If it shares our DNA, then perhaps each of us would have niches on the planet to exist side by side. If it is the product of a unique genesis, we would have to prevent cross contamination. If there's no preexisting life, then we could consider introducing the Terran genome.

If any ecosynthesis is to be done, Robinson proposes that we should proceed only up to a certain elevation in the terrain, then stop. Don't create an atmosphere to cover the entire planet, but only part of it. In *Green Mars,* where he frames what will become the basis for a Martian constitution, it's one of the document's principal tenets:

> The Martian landscape itself has certain "rights of place" which must be honored. The goal of our environmental alterations should therefore be minimalist and ecopoetic, reflecting the values of the areophany. It is suggested that the goal of our environmental alterations be to make only that portion of Mars lower than the four-kilometer contour human-viable. Higher elevations, constituting some thirty percent of the planet, would then remain in something resembling their primeval conditions, existing as natural wilderness zones.

There's a nineteen-mile difference between the lowest point on Mars in the giant Hellas Planitia—a crater 1,250 miles wide with a floor 3.7 miles below the mean level of the Martian surface—and the highest point along the crater rim of Olympus Mons. Robinson, by proposing to foster an atmosphere only for the bottom two and a half miles of that vertical relief, would create a wilderness preserve defined by elevation, a situation somewhat analogous to where he lives in Davis, California. The Central Valley is farmed or built over on every square inch of its floor, and towns have crept up all along the foothills of the Sierra. But at one contour line or another, depending on the specific terrain, the slope and other factors make it not cost-effective to continue building, which is traditionally where public lands begin. In some cases, where the scenery is considered sublime, there's a national park or wilderness area, but in general environmental policy has been dictated by economic forces. Robinson is arguing that on Mars we should set the policy first.

An even more interesting idea that allows humans to wander portions of Mars unimpeded while maintaining the natural state of its surface is Greg Bear's proposal that we modify the people instead of the planet, to genetically engineer a branch of the human race into becoming Martians, a concept Pascal finds especially appealing. It has its own set of ethical issues, obviously—creating a race that could live in a low-pressure environment of carbon dioxide, but would thus be incapable of living on Earth, for example. But the idea of taking the responsibility for transforming ourselves versus the environment has a certain moral elegance. And as Bear points out, if it could be done at all, it would be cheaper and faster.

Pascal agrees, noting that genetic engineering seems to be progressing more quickly than viable terraforming technologies, and he admits to being a fan of the premise underlying *Star Trek*'s famous prime directive about not making your presence known on planets where life is less evolved than on ours: Don't mess with the course of history. He extends that down to the microbial level, thinking it acceptable to study them if you don't eradicate the tiny creatures. But he is also a pragmatist who has read enough history to know that such ethical considerations may hold during the exploration phase, but when humans go beyond that toward colonization, the urge to conquer and control is very difficult to overcome.

An argument made by many people about why we should explore Mars and other planets is that it ultimately will give us more information about Earth, and provide us with a clearer picture of how vulnerable our planetary habitat is to human presence. That's a valid argument that Charlie Cockell backs up at length in his newest book, *Space on Earth,* which is about "environmentalism and the space-faring civilization." We so far have only one planet—our own—as a baseline in our understanding of how Earth's planetary ecology works. The more we understand the dynamics of other planets, the better chance we have of understanding our own.

I would submit that the arguments we use for preserving wilderness areas on Earth as ecoreserves apply to other worlds as well. Preserving intact preexisting ecosystems on other planets would preserve biological diversity just as it does on Earth, but on a larger scale. Robinson and Bear are thinking about ways to accommodate both nature and life, both human and extraterrestrial rights and needs. Chris McKay has argued that because there are other bodies in the solar system that we don't have a remote chance of modifying—Mercury, the moons of Jupiter, and so forth—we can afford to change Mars. I would counter that Mars may be the only place where we have the opportunity to live in an alien environment, and at least some of it should be preserved.

Even if Mars is sterile, maintaining a significant portion of the Martian environment as it is would help us to better know ourselves. I don't mean that in any abstract psychological way, but in a very specific and concrete neurophysiological sense. Just as we are able to see more clearly how we process the world visually when we are in a desert environment—a place where we are poorly adapted to survive, hence the inadequacies of our visual systems are more apparent—so, too, would an experience on Mars allow us to measure an even fuller array of cognitive responses as a result of being on a planet where life did not evolve and where the terrain offers us multiple challenges just to stay alive, much less settle or prosper. That could not fail to give us new perspectives on how we perceive Terran landscape, thus shape it with our tools.

Pareidolia, that cognitive habit of assembling random or ambiguous stimuli into a known pattern, be it a predator lurking in the foliage or a face on Mars, produces more than simple misperceptions and clues to how our neurophysiology functions. It also projects what is literally on our minds, and is the basis for the Rorschach inkblot test and other associative psychological experiments. Alien nature, even without life, provides us a wider array of opportunities to measure ourselves. Keeping Mars alien will provide us not only a wider, deeper field of scientific opportunity, but a place in which to see ourselves as alien. That can't be done on Earth; it can only be done somewhere else.

We're well on our way to homogenizing Earth, to making it a landscape that is the home of the same monolithic race and culture. It behooves us to preserve the very differences that we are built to perceive in nature, lest we lose sight of both the marvels and the dangers that are the universe. Which is to say, lest we lose sight of ourselves.

CHAPTER TWELVE
Of Sled Dogs, Seals, and Tourists

EARLY MORNING in Resolute the next day, while decompressing from Devon and before leaving what is still a terrain with some discernible difference from the rest of North America, I walk out along the shoreline to the west in a dense cold fog. I want to revisit an excavated Thule winter village, one of four such sites near the town. To my left, where Paul and I had ridden on the sea ice, the bay is mostly ice free. That means that the polar bears who frequent the local dump nearby won't return until winter—but I'm still feeling a bit exposed without a shotgun. It gives me some relief to see that many of the village dogs are staked out along the shore, where they live most of the year. Were there a bear around, they'd let me know.

After almost an hour of walking I reach the archeological site, where a modest sign notes that people lived in its several houses around 1400 AD. The scattered one-room habitations are sunken partway down into the south-facing sides of hummocks and floored with stones. Raised benches for sitting and sleeping surround the hearths. The houses are so small that, were I to stretch out in one, my fingers and toes would span the entire floor. White whalebone beams like those that once supported the roofs of sod arc over the remains to meet in the center, making the village a collection of skeletal monuments.

It's a desolate scene even in summer, and I can't imagine how isolated the dozen or so families that lived here must have felt during the long winters with their twenty hours of darkness every day. The Inuit name for Resolute is Qausuittuq, which means "the place with no dawn."

By 1650 the climate had changed once again, the bowhead whales that the Thule hunted were gone, and they had switched to living on walrus and the occasional caribou and fox. When Parry's expedition arrived in 1819, apparently the first European contact with the islanders, the few remaining people begged for food. Life on Cornwallis has always been intermittent, and today some of the villagers worry about Resolute disappearing with the closure of the Polaris lead mine. Tourism remains the only likely growth industry on the island, mainly ecotours to Devon Island, where people pay handsomely to visit one of the most pristine coastal regions left in the world.

The truth is that Cornwallis is inhabited only by a fluke of geopolitics, and the temporary presence of a strategic metal that is now played out. The weather station and airstrip, the logistical support provided by Polar Shelf, even the hotels for the tourists are all only temporary fixtures dependent on the vagaries of climate change. As it gets warmer and the Northwest Passage opens up, the town might flourish—or if Paul finds his diamonds on Devon. If it someday becomes colder, the passage ices up, and flying conditions became more difficult, the Inuit might have to relocate once more, as they have throughout the history of habitation on these islands. What is now the Canadian High Arctic was explored thousands of years ago because people were looking for new terrain that was available and clement enough to become territory, a place to live and hunt and raise families. It was settled because, at the time, it was both feasible and all of the more hospitable lands to the south were already occupied. If given options to settle in the boreal forest, people might have chosen differently.

It's a telling comment on modern life that we choose to visit extreme environments such as the High Arctic and the Antarctic, and in some cases want to work and live in them. We seek out the edges of existence as far from the clutter of a consumerist society as we can manage. We seek to escape a deluge of advertising, the thousands of messages we receive daily that so overlap one another that the cascade becomes white noise. We hope to locate a less mediated and more authentic version of existence. If we are not personally exploring for new terrain to settle, we remain nostalgic for that activity, one that we feel to be a noble and essential endeavor. Operators of adventure tourism companies manage to make a living off precisely that atavistic drive.

After Kim Stanley Robinson wrote the Mars books, he went to the Antarctic as a visiting writer with the National Science Foundation. The near-future science fiction novel he based on the experience, *Antarctica,* not only has tourists on the ice re-creating the adventures of Shackleton and Scott, but also posits the existence of a group of "feral" Antarcticans who manage to live on their own and mostly outside of both conventional society and the U.S. government's logistical umbrella. The fictional conflict is between people who attempt to create a sustainable social ecology on the continent and members of a military-industrial complex who wish to exploit it for mineral development. In a 1997 interview in *Locus* magazine, Robinson described it this way:

> My original idea for *Antarctica* was to have some kind of conflict between development forces interested in the natural resources down there, especially oil and coal, and the kind of wilderness park idea that has been proposed for Antarctica,

> of completely leaving it alone. What interested me when I got down there was the idea of some third way that was not just exploitation of the natural resources and also was not leaving it alone as a blank wilderness, but was an attempt at a human habitation of Antarctica that went beyond the scientists' visits that are happening there now—as a science fiction act of extrapolating a possible future society.

His ferals fish for Antarctic cod and eat seal steaks, krill cakes, and carefully culled penguin eggs. They raise their produce in camouflaged greenhouses, which they supplement with clandestine groceries flown in from New Zealand. Essential equipment is salvaged from abandoned bases or smuggled in during the summers. The novel combines Robinson's vision for Mars with Arctic cultures both historical and contemporary, an updated version of the life that had been attempted here in these stone huts dug into the permafrost.

The Antarctic, however, is an even fiercer environment than the Arctic. It can get as cold as minus 145°F, its winds blow at more than two hundred miles per hour, and it is the most isolated continent on Earth. Steve Pyne once analyzed the budget of the NSF and calculated that it cost $10,000 per day to support a single person on the ice. Yet it's a picnic compared to Mars, which is not only much colder and bombarded with lethal radiation, but its very surface is toxic with peroxides, the dust from which invades everything. Given that science and engineering march onward in always surprising ways, it's perhaps not so much an issue of whether we could surmount the technical obstacles of getting people to Mars and supporting a small colony under such conditions while there—it's just that we might not be able to afford it. People on Mars, like both the Inuit of Resolute and the scientists at the South Pole, would be dependent on support from Earth during any foreseeable future, which means the cost-to-benefit ratios would be upside down.

Although Robinson never loses sight of the fact that his utopian societies in the Antarctic and Mars depend upon support, however clandestine and indirect, from urban societies on Earth, his fictions allow the economics to work. His characters not only choose to live in these difficult environments, but also have the luxury of trying to figure out their preferred politics while doing so. What he is acknowledging, macroeconomic issues aside, is that it's a hardwired instinct for some of us to sit out on the edge of any available terrain, and that we will probably attempt to settle Mars. The usual pattern is that after the explorers and scientists come the tourists and settlers, the latter two groups encouraged by images that show the new spaces being converted into places. Based on what I've seen in the more extreme environments on Earth, however, I wonder if it might not end up that scientists and tourists will be more likely to go to Mars than colonists, given that the environment and logistics for Mars are much more like those of the Antarctic than the Arctic.

The layers of Arctic culture, although they have been accumulating for millennia, are relatively few and thin compared to those of more temperate latitudes for obvious reasons, the cultural ecology paralleling that of the natural world. The Arctic is relatively unpopulated, hence still a place of mystery and allure for most people. Its landscape and natives have not been fully processed, assimilated, or understood by most other peoples. In common parlance, they have not been spoiled or tainted by commerce, and thus retain a degree of metaphorical purity in the minds of most tourists. To visit the Arctic is supposedly to approach more closely the edges of geography and even of history. The culture of the Antarctic, because it has no indigenous people, is infinitesimally thin by comparison, and in those terms remains an even more alluring destination. In a continent the size of the United

States and Canada combined, there are as few as three thousand people working there during the austral summer, and no permanent residents outside of fiction.

Robinson's Antarctic novel, along with a handful of other fictions, the various expedition accounts, photography books, films, and the paintings, poems, and sculptures created by the fewer than two hundred artists, writers, and filmmakers who have visited the continent, are the sum total of the Antarctic arts culture. Their works are all that make the Antarctic a place in the minds of most people, not simply a space. Mars, despite the panoramas sent back by the rovers and all the satellite images, is at this point little more than a map with fanciful illustrations drawn around the borders. It's no wonder that the arguments about planetary protection protocols are debated at such length within the space community. Not only is an ecosystem perhaps at stake, so is its metaphorical purity—meaning our access to the edge of human experience.

More and more people are flocking to the Arctic and the Antarctic as tourists, eager to see for themselves what only explorers and scientists and a few adventurers have experienced previously. Resolute already has three hotels; if it ever becomes feasible to build a hotel on the Antarctic Peninsula, the banana belt of the continent, I have no doubt someone will try it. The development of tourism in the Nepalese Himalaya provides an example of how the process works.

The rugged southern slopes of the world's tallest mountains were closed to climbers for much of the twentieth century, opening up only after the Chinese invaded Tibet in 1950. That made it strategic for India to encourage the strengthening of Nepal as a buffer zone between itself and the monolithic Red Army pressing down from the north. As a result, Nepal began to offer climbing permits to the peaks along the Tibetan border, Mount Everest included, and Sir Edmund Hillary and Tenzing Norgay became the first people to summit in 1953. The subsequent publicity, much of it photographic, brought more climbers, and then some of the more adventurous travelers. Ten years later the first Americans summited with an expedition sponsored by *National Geographic,* and the Sierra Club published a lavishly illustrated book about the climb in 1966. Lute Jerstad, one of the team members, was running treks into the country by 1970, at which point the Nepalese economy in the Everest region began to shift from traditional agrarianism to tourism.

Shortly thereafter a Japanese company opened the Everest View Hotel above Namche Bazaar, a major village along the route to the peak. At 12,779 feet it is still the highest hotel in the world, and when the weather is good you can spy the plume flying from the top of Mount Everest where it pokes into the jet stream. Despite the fact that fresh oxygen is pumped into every room, visitors who have failed to acclimatize properly have died from altitude sickness. That doesn't stop people from paying more than $300 a night to stay there in order to say that they have seen for themselves the pinnacle on Earth that is closest to outer space.

I worked as a trekking guide in Nepal briefly during the 1970s and late 1980s, and most of my clients had been lured to the region by color photographs in books and magazines, exploration and adventure narratives, and a fervent desire to see the highest point on the planet, a global landmark. The images of the time for the most part depicted a country of modest villages tucked discreetly on the lower slopes of the mountains, and smiling Sherpas wearing traditional Tibetan clothes. Trekkers wanted to take pictures, enjoy the fresh air and an ancient exotic culture, and take more pictures.

The Antarctic has long since replaced Nepal as the premiere tourist goal on the planet, and it is fairly predictable that many of us who earlier worked in the Himalaya would want to go there, including Kim

Stanley Robinson, who set his first novel in Kathmandu after a visit to the Asian kingdom. It is also no surprise that some people who have worked for the various national polar programs have later become guides in the Arctic and Antarctic, working on everything from cruise ships to mountaineering expeditions, their clients once again lured by the ongoing stream of color photographs, the predominate visual medium through which the regions are represented to the culture at large. The words "pristine" and "pure" often appear in the captions.

The ultimate tourist trip is currently, of course, a visit to the International Space Station, which only two people have managed so far at a cost of $20 million each. Tourism to Nepal grew from virtually nothing in the late 1960s to more than thirty-eight thousand people per year at its peak in May of 1999 (just prior to the increasing Maoist violence, which has sent the industry there into a tailspin). In Antarctica it grew from two thousand people in the late 1980s to more than twenty thousand annually in 2005. Space tourism, despite the expense, will also grow. NASA and other space agencies, notably the Russian and the Japanese, have already studied low-orbit tourism as a way to support their other ventures in space. No doubt ex-astronauts will eventually serve as guides in space.

I don't believe the attraction for people traveling to Nepal, the Arctic, or the Antarctic is simple thrill seeking or bragging rights, but a genuine desire to remember what it is like to be an explorer and to see the shape of where we live. It's that drive to get to the edge of the world, where we feel as if we can hold the entire planet in our vision—and to imagine what it might be like to go even farther.

It's not wise, however, to push too far the comparison of how we perceive Mars as a Rorschach test for human society. For one thing, inkblot interpretation is a notoriously subjective procedure. But it's a useful metaphor for the process of how we treat what we perceive to be empty ground. We project our own personalities upon it, our "wishes, lies, and dreams" as the poet Kenneth Koch once titled a book. It's been a useful survival trait for humans to be able to first imagine ourselves in a new place, and then fulfill the dream of settling that new terrain into territory. But as Pyne reminds us, it's come at a cost. Perhaps we are beginning to move beyond that stage in exploration.

We have the opportunity to explore Mars as not just a matter of what we *can* do, but also what we *should* do. The former is a matter to be determined by scientists and engineers. The latter is something we might conceive of as a matrix of possibilities, options to be imagined by people who will understand Mars not just as a space to be exploited, but a place to be considered.

Precarious, I say to myself as I turn away from the tiny houses in the Thule village. Very precarious. On my walk back into town I swing by the wreckage of the jetliner at the end of the airstrip. The thin aluminum skin of the fuselage looks like torn paper. The Arctic towns and Antarctic bases are surrounded by the debris of crashed aircraft, reminders of how difficult it is to support human life in a hostile environment out at the end of a long logistical pipeline.

I huddle and shiver in my parka whenever I'm not moving, so I leave the debris behind. What would Cornwallis Island be like, I wonder, if the Inuit families hadn't been resettled here in the 1950s? Perhaps not so much different than if Resolute becomes too expensive to support with groceries flown in on jet airplanes, and families subsidized with government funds in hopes of maintaining sovereignty over the

Northwest Passage and the High Arctic. The buildings would become buried ruins, a site for future archeologists to excavate and label next to the Thule villages.

Working on Devon Island has led me to look not only at the business of what we can do to get to Mars, but at the factors we might consider when we think about what we should do once we get there. It's a good idea to practice driving around on Mars so that we don't send a human expedition that is dependent upon an untested new technology—such as the tinned food that helped bring Franklin's expedition to disaster. But that's just part of the engineering challenges. Looking at the history of exploration, science, development, and tourism on Devon and Cornwallis—that can give us clues about what we should or shouldn't do.

As I walk back into town, I pass by the dogs again. The earliest remains of domesticated dogs, which probably evolved from wolves in Asia, date from between twelve thousand to fifteen thousand years ago, a period when they were already in the Siberian Arctic. Dogs have been prominent companions to people in the North American Arctic for at least the last four thousand years, and early attempts to reach the North Pole used them for pulling sledges. Working dogs made their first appearance in the Antarctic in 1898. Amundsen used eighty-six of them in 1911 to pull his way to the South Pole, then slaughtered dozens of them for food on the return journey, an idea he adopted from Arctic practices. Dogs worked well in the Antarctic, and were used extensively by various national programs there up through the 1960s, when snow machines replaced them. Nonetheless, they were kept on the ice for travel through crevassed areas, where they were slower but safer, and for the pure pleasure of their company. Seals were shot to feed them, a practice not everyone approved of, but one outweighed by the psychological benefit of having the dogs around during the long winter months of darkness.

Then it was discovered that the dogs were transmitting canine distemper to the native seals, and the Antarctic Treaty of 1993 banned them. The dogs were shipped home, despite their usefulness and their popularity with the personnel. That's quite a contrast to the dogs that the residents of Resolute keep chained here along the seashore. But then, seals in the Antarctic aren't available for hunting, while those in the Arctic are. Part of the reason is that there is an indigenous tradition of consuming seals in the north for subsistence, whereas sealing in the south was done almost exclusively for commercial reasons, there being no native Antarcticans. In turn, those realities have influenced subsequent images made of seals—they have become part of the sacrosanct Antarctic landscape.

Analog Mars-on-Earth environments aren't just places where we can practice driving around Mars with some degree of verisimilitude. They are also places where we can explore the limits of our cognition, our perceptions, and hence our values. Pictures of the Antarctic landscape and wildlife—both framed, literally and metaphorically, as unmarked by humans—have been critical to how we have shaped our conservation ideas. Examining that process in extreme environments, where it is easily observed, allows us to understand how and why we form our ideas of space and place the way we do.

We're already well on our way in developing our ideas about Mars, as the two camps on Devon demonstrate. If I had to place bets on how we will end up treating Mars, I think it will be more like the Antarctic than the Arctic, although I find Devon Island a better environment than the Antarctic in which to test exploration protocols for Mars. It's mostly land, not ice, for one thing, and it offers a more comparable surface upon which to drive. But in terms of environmental severity and isolation, the Antarctic is a closer physical analog for Mars. I believe it will prove to be a closer cultural one as well, and it would not be surprising to see the planet placed under the protection of a multinational treaty based on the one that protects the Antarctic—which is not a country, but a continent.

Mars may shift from an unknown space to a place in our imaginations, but not be a terrain that we convert to territory. It may not become real estate and a source of mineral wealth for Earth. It may remain a frontier for science that we never close, an *Ares incognita,* an alien place in which we measure our minds.

AFTERWORD

JANUARY 2006, and Devon Island is locked up in winter stasis. It's completely dark, thirty below zero, and the weather uncertain. The greenhouse is cold and a set of polar bear tracks circles the kitchen tent, soon to be erased by the wind. On Mars, however, the season has already started to shift toward spring. The rovers that Steve Squyres and his team had planned to work for only ninety days have survived an entire Martian year—which is to say more than seven hundred days, or almost two Earth years.

In early June of 2004 Squyres and the mission planners had made the decision to send *Opportunity* down into Endurance Crater. At first the slope was no more than twenty degrees. The engineers at the Jet Propulsion Laboratory in Southern California built a test ramp that had a slope of up to twenty-five degrees, and the spare rover they ran down it handled it just fine. By midmonth *Opportunity* was sixteen feet down inside the crater and Squyres, who had been expecting to see volcanic basalt, was looking at still more sulfates, meaning that salty water had existed even longer than thought on the planet. The decision was made to continue downward. By the end of July the rover was forty-three feet down in the crater, had driven almost a mile during the 170 Martian sols, and was taking long sleep breaks to conserve energy in the deepening winter chill. On the opposite side of the planet, the twin rover, *Spirit,* was starting up the Columbia Hills, a series of seven small summits named for the astronauts lost in the *Columbia* shuttle explosion.

Just before the arrival of 2005, *Opportunity* emerged from inside the depression where it had spent the last six months. It sent back pictures of high clouds, and there was even a trace of frost on its antennae, evidence of the increased moisture in the thin atmosphere as the Martian spring approached. *Opportunity* had crept down slopes with more than thirty degrees of incline. In places its wheels could find no traction at all, but the drivers at JPL were patient, each time finding a way to maneuver gently out of the impasse. The rover had gotten as far eastward in the crater as it could safely go—within fifty feet of its goal, the Burns Cliff—before the terrain became too sandy and impassable. For several days it trained its cameras and then thermal emission spectrometer on the thirty-three-foot-high outcrop of layered bedrock. Squyres, looking at the images, reported that it appeared as if some of the lower portions had been deposited by wind, indicating that water had been shallow and intermittent at the site. When I saw the wiry young scientist speak at the NASA Astrobiology Science Conference that year, he repeated what had become his mantra: "Follow the water." The two rovers had found evidence of water almost everywhere they looked.

While *Opportunity* was inching around the crater slopes and *Spirit* was examining outcrops in the hills, the European Space Agency's *Mars Express* was orbiting overhead and measuring atmospheric gases. It confirmed what a telescope in Hawaii had observed earlier in the year: the presence of methane. The gas is the simplest hydrocarbon there is, just one carbon and four hydrogen atoms that come apart when encountering ultraviolet light. For any methane in the Martian atmosphere to be observed today, it must have been put there no longer than three hundred years ago. Save a herd of cows wandering through the Valles Marineris, there are only two mechanisms of which we are aware that could account for this. One is geothermal activity, which means volcanism somewhere on or near the surface of the planet. The other is that class of bacteria (methanogens) that breathes out the stuff.

Until 1999 no one would have been able to offer any evidence of volcanoes erupting on Mars during the last five or six hundred million years. Then that year scientists saw evidence of lava flows as young as perhaps forty million years. In the waning days of 2004 a series of high-resolution stereo photographs taken by the European orbiter showed smooth and uncratered lava flows on some of the shield volcanoes. Statistically, those areas should have shown just as much evidence of impacts as the surrounding areas. They didn't. There may have been volcanic activity on their slopes as recently as two million years ago, which is within the last 1 percent of the total time Mars has been a planet. And at the base of Olympus Mons there appeared to be glacial deposits laid down only four million years ago. That's considerably younger than some of the ice in the Antarctic.

All of these observations were single points of evidence, and it was very tempting to link them together in that exercise of intellectual pareidolia we call speculation. The relatively young flows of lava meant that there could still be volcanic activity on the planet that we hadn't spotted yet, and those volcanoes could be producing methane. The ice meant water is on or near the surface. Ice and heat can produce wet environments conducive to life. On the other hand, if no volcanoes are fueling geothermal vents and still releasing methane, that leaves bacteria as a primary suspect for the gas.

To determine if this is a line of reasoning more solid than Lowell's canals, many more points of data will need to be collected. The next mission slated is the *Mars Reconnaissance* orbiter, which in 2007 will start to increase the resolution of our surface images from the scale of a dining room table down to serving platters. The next year we'll send up a lander to the northern polar region of Mars to look for organic compounds in the ice, but it's not until 2010 that another rover will land on the planet, the *Mars Science Laboratory*. This larger and more complicated version of the current rovers may be able to roam farther and in rougher terrain, and to actually search for organics in the regolith.

These plans are logical but small steps, and at the present rate it could be another thirty years before humans are sent to explore Mars in person. In his office at Moffett Field, Pascal is slotted behind a desk surrounded on three sides by overflowing bookcases and piles of recent NASA reports, observational data, and scientific papers. He taps away on his laptop, filling out the details of what he calls "Mars Indirect," an idea for a mission that would get people to Mars—or at least the very near vicinity—sooner rather than later.

The largest single expense for a manned mission to Mars is getting down and then, even more critically, getting back up the gravity well of the planet. That's what would take the most amount of fuel, hence money. The technology already exists to transport that much mass, but it makes contemplating a manned trip prohibitively expensive in terms of political will, given the country's current fiscal picture. Pascal has the idea that some of the science to be done on the surface of Mars could be done on the surface of Phobos, the larger of its two tiny moons.

In approximate terms, Phobos measures 16.2 by 12.9 by 11.4 miles, and it circles over the Mars equator only 5,627 miles above the center of the planet. Its proximity to the planet below means two things. A mission to Phobos would put scientists on a planetoid covered with remnants of the Martian surface blasted off by impacts and sent into space—the same stuff that produced the Martian meteorite in the Antarctic that may contain fossilized nanobacteria. In fact, ancient Martian fragments on Phobos would be more perfectly preserved than even the rocks on Mars, not having been subjected for millennia to the chemicals and weathering on its surface. And, it would be an ideal platform for a science base from which to observe the surface below.

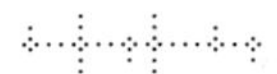

As 2005 edged into the New Year on Earth, Squyres and his colleagues celebrated the fact that the rovers had survived for a complete Martian year. Each of the rovers had traveled more than three miles, and together sent back almost 130,000 images. *Opportunity* was on its way to a much larger crater a half mile in diameter, and *Spirit* had become the first envoy of humankind to climb a mountain on another planet. In several remarkable series of images, dust devils were photographed traversing the landscape. Sitting at JPL with Doug Alexander and Bob Deen to watch one of the midday microtornadoes, I felt the hair rise on the back of my neck as we watched the dusty atmosphere actually moving on Mars. The images of the dust devils were stitched together from a quick series of still shots, and to see something moving on Mars has become a favorite activity for people at JPL. The wind events also turned out to be useful: They periodically cleaned off the solar panels, and the rovers were maintaining close to full power capacity.

Alexander is the team lead for processing images from the rovers, and Deen is the chief software engineer whose algorithms make it possible. Their domain is the Multimission Image Processing Lab on the seventh floor of JPL's Space Flight Operations Facility, a building that for all its history and glory looks more like an older suburban office building than the forefront of human exploration. The room is quiet and dim with various workstations and computer monitors partitioned off from one another. Every image received from both rovers has to be analyzed through up to twenty-five separate filters, and the work has to be done quickly, as the information must be given within hours to the mission planning team and drivers so they can navigate the next few feet. Although Alexander and Deen have long since been off the Martian sol sleeping schedule that everyone worked during the first few months of the mission—which put them off Earth time by an additional forty minutes each day—they still lose track of where in the week they are and what hour it is outside. But in addition to processing the images for JPL and NASA personnel, they've also been able to put up more than 110,000 images online, and there are now thousands of people worldwide who virtually go to Mars every day. Among the more popular downloads are the dust devil sequences.

Almost as impressive and immersive as watching the "moving" images were the large panoramas hung in the hallways. While *Spirit* was climbing up the Columbia Hills, the highest summit a thousand feet above the bottom of the Gusev Crater, it captured a 360-degree panorama that looked back along its tracks and down to the plain below. It's one of two hundred such wide-view pictures assembled from *Spirit* images, and when you come across one printed out along several yards and mounted at eye level, it's almost like looking out a window onto the surface of the planet. You feel as if you could reach out and touch the sand dunes below the rover.

The rovers are, of course, wearing out, and the cumulative strain on Squyres and other team members is apparent. Not only are they busy cataloging fresh discoveries every week, such as new kinds of rocks not seen before on Mars, but they're having trouble coming up with names for all the new features the rovers are encountering. Every rock, outcrop, and depression needs a reference name, and the scientists have resorted to borrowing nomenclature from ancient Mayan cities, pop music groups, and even individual bones in the human body. They've also appropriated names from Arctic and Antarctic exploration. How long, I wonder, before they start using Inuit names? That would be a nice symmetry, an implicit acknowledgment that we have been overlaying names from our space program across features on Devon Island that had already been named by the Inuit.

Despite the amount and complexity of data that the rovers have generated, they have opened more questions than they've answered. No one can figure out, for example, how much water has been on the surface of the planet: shallow lakes that came and went within centuries—or shallow seas that lasted for millennia? Some planetary scientists have proposed that features seen by *Opportunity* near its landing site weren't created by water at all, but by massive explosions caused by multiple meteorite strikes. Others have proposed that evidence taken by Squyres to mean water was left behind by volcanic steam. The rovers won't be able to solve these issues definitively, nor will Pascal's proposed mission to Phobos. To comprehend areology, we're going to have to go to Mars in person.

One of the few unambiguous sights captured by *Opportunity* this year, as it continued to creep forward at two inches per second, was a human artifact on Mars, the crumpled remains of the rover's own heat shield. The cone, which had been shed ninety seconds before the rover touched down, had landed about a half mile from where *Opportunity* finally came to rest after bouncing across the surface. When the rover climbed out of Endurance Crater, the shield was only 220 yards away, and NASA decided it would be interesting to examine it.

Squyres noted in a press conference that there were two reasons to visit what was once an expensive dish of honeycombed metal and exotic materials and was now space junk. The scientists wanted to see what kind of impact it made on the surface—a deeper hole than what the rover could dig with its wheels—and therefore what was revealed by the relatively new minicrater. And second, the engineers who designed the shield needed to see how it had weathered the entry into the Martian atmosphere, when it slowed the spacecraft down from twelve thousand miles per hour to one thousand miles per hour within the space of only a single minute, and reached temperatures of more than 2,600°F. (Needless to say, the defense contractor who fabricated the shield, and who builds military missiles, also had a vested interest in knowing the results.)

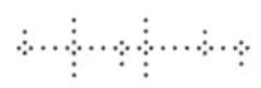

As the rover approached, it sent back pictures of the 198-pound pyramid of aluminum sitting on the sandy plain, a surface rumpled by wind that stretches unhindered toward the horizon. The shield was crumpled down to half its original diameter, and it became apparent that it literally had been turned inside out by the landing. This proved a challenge to engineers attempting a thorough examination of the surfaces exposed to all the heat. But it had done its job and will remain on Mars intact for centuries, or until buried by windblown sand.

A third and unstated reason for *Opportunity* to pause by the remnant was to provoke self-reflection in us, and provide a chance for us to see our own presence within the Marscape. The heat shield is one of only a handful of artifacts on the planet, which means that from the moment it plowed into the surface it became a historical relic, a technological meteorite that memorializes a literal trial by fire on an alien world. Just as we use Devon Island as an analog environment for Mars, so the rover and heat shield form an analog for Earth on Mars. The shield stands in contrast to the wilderness around it, a reminder of what life is and how powerfully it seeks to drive itself from planet to planet, no matter the peril.

ACKNOWLEDGMENTS

MY PROFOUND THANKS are due first to Pascal Lee, who invited me to the National Aeronautics and Space Administration's Haughton-Mars Project on Devon Island in the Canadian High Arctic—and then to my friends and co-workers on Devon Island during 2002 and 2003: Rhoda Akeeagok, Paul Amagoalik, Joe Amarualik, Matt Bamsey, Adrien Bisset, Keegan Boyd, Stephen Braham, Sam Burbank, Charles Cockell, Keith Cowing, Ray Demblewski, Rainer Effenhauser, Trish Garner, Brian Glass, Thomas Graham, Steve Hoffman, Melanie Howell, Corey Ippolito, Jeffrey A. Jones, Greg Kalluk, Jamie Kalluk, Jeffrey Kheraj, Stan Kusmider, Darlene Lim, Carlos Nieto, Addy Overbeeke, Gordon "Oz" Ozinski, John Parnell, Sandy Salluviniq, John W. Schutt, Matt Silver, Robert Stewart, Elaine Walker, and Larry Young. I also owe gratitude to the pilots of Kenn Borek flying out of Resolute Bay, and the fearless canine Kimmiq (owner: Simon Idlout, Resolute Bay), who kept the camp safe from polar bears. The Mars Institute and its director, Marc Boucher, were critical sponsors of my visits to Devon Island.

Guy Webster at the California Institute of Technology's Jet Propulsion Laboratory took me over to the Space Flight Operations Facility and introduced me to Doug Alexander and Bob Deen in the Multimission Image Processing Lab in order that I might watch how images of Mars are assembled from the data sent back by the Mars Exploration rovers.

Fellowships from the John Simon Guggenheim Memorial Foundation and the National Endowment for the Humanities funded much of the travel and writing. I am obliged by the grant conditions of the latter to note as follows: Any views, findings, conclusions, or recommendations expressed in *Driving to Mars* do not necessarily reflect those of the National Endowment for the Humanities.

Additional time for research and writing was provided by two residencies. The first was with the Lannan Foundation's writers-in-residence program in Marfa, Texas, during the fall of 2002—my appreciation to Martha Jessup and Douglas Humble. The second was with the University of Nevada, Reno, as a visiting Hilliard Scholar a year later, arranged by Bob Blesse, Cheryll Glotfelty, and Scott Slovic, all of whom are long-standing colleagues. In addition, I started preparing for my travels while a visiting scholar at the Getty Research Institute in Los Angeles; my gratitude to the GRI and library staff—in particular Thomas Crow, Charles Salas, Sabine Schlosser, and Jay Gam—is beyond words.

Driving to Mars is the first manifestation of what I hope to be a long collaboration among myself, my agent Victoria Shoemaker, and a publisher whom I have long admired, Jack Shoemaker. Working with Jack and others at Shoemaker & Hoard is an exceptional experience, and I gleefully extend my appreciation to Roxy Font, Michelle Gray, and their now independent colleague Heather McLeod, who edited the manuscript.

Kim Stanley Robinson remains an inspiring friend throughout my travels; his intellect and talent make them richer experiences by far, and improve my own work.

I am forever grateful for the support of my friends David Abel, Robert Beckmann, Ric Hardman and Mickey Guston Hardman, Jeff Kelley and Hung Liu, Joe Mastroianni, and Enrico Martignoni. It would have been impossible for me to do this work without their help, and they have my deepest gratitude.

Finally, my friend and partner Karen Smith not only kept me in e-mails while I was in the Arctic but bought me the chair in which I wrote the book.

BIBLIOGRAPHY

THE ARCTIC

Beattie, Owen, and John Geiger. *Frozen in Time: The Fate of the Franklin Expedition*. Vancouver: Graystone Books, 1987.

Dowdeswell, Julian, and Michael Hambrey. *Islands of the Arctic*. Cambridge: Cambridge University Press, 2002.

Hallendy, Norman. *Inuksuit: Silent Messengers of the Arctic*. Vancouver: Douglas & McIntyre, 2000.

Kane, Elisha Kent. *Arctic Explorations in the Years 1853, '54, '55*. Philadelphia: Childs & Peterson, 1856.

Karpoff, Jonathan M. "Public Versus Private Initiative in Arctic Exploration: The Effects of Incentives and Organizational Structure." *Journal of Political Economy* 109, no. 1 (February 2001): 38–78.

McGhee, Robert. *Ancient People of the Arctic*. Vancouver: University of British Columbia Press, 2001.

Mowat, Farley. *High Latitudes: An Arctic Journey*. South Royalton, VT: Steerforth Press, 2002.

Pielou, E. C. *A Naturalist's Guide to the Arctic*. Chicago: University of Chicago Press, 1994.

Riewe, Rick, ed. *Nunavut Atlas*. Edmonton: Circumpolar Institute, 1992.

Ross, John. *Narrative of a Second Voyage in Search of a Northwest Passage and of a Residence in the Arctic Regions, During the Years 1829, 1830, 1831, 1832, 1833*. London: A. Webster, 1835.

Ross, M. J. *Polar Pioneers*. Montreal: McGill-Queen's University Press, 1994.

Savours, Ann. *The Search for the Northwest Passage*. New York: St. Martin's Press, 1999.

NASA EXPLORATION/MARS

Bergreen, Laurence. *Voyage to Mars: NASA's Search for Life Beyond Earth*. New York: Penguin Putnam, 2000.

Boyce, Joseph M. *The Smithsonian Book of Mars*. Washington, DC: Smithsonian Institution Press, 2002.

Caiden, Martin, Jay Barbree, and Susan Wright. *Destination Mars: In Art, Myth, and Science*. New York: Penguin Putnam, 1997.

Clancey, William J. *A Framework for Analog Studies of Mars Surface Operations*. Moffet Field, CA: NASA/Ames Research Center, 2000.

———. *Human Exploration Ethnography of the Haughton-Mars Project, 1998–99*. Moffett Field, CA: NASA/Ames Research Center, 2000.

Cockell, Charles. "From the Cradle to the Stars: Environmentalism and the Space-Faring Civilization." Working manuscript, 2004.

———. *Impossible Extinction: Natural Catastrophes and the Supremacy of the Microbial World*. Cambridge: Cambridge University Press, 2003.

———. "Mars Is an Awful Place to Live." *Interdisciplinary Science Reviews* 27, no. 1 (2002): 32–37.

———, ed. *Martian Expedition Planning*. Science and Technology Series 107. San Diego: American Astronautical Society, 2004. Proceedings of the Martian Expedition Planning Symposium of the British Interplanetary Society held in 2004. This compendium includes numerous papers by participants in both the NASA Haughton-Mars Project and the Mars Society Arctic and Desert Mars Stations (FMARS and DMRS).

Cowing, Keith. "Earth on Mars." *Ad Astra* (May/June 2002): 18–21.

Fogg, Martyn J. *Terraforming: Engineering Planetary Environments.* Warrendale, PA: Society of Automotive Engineers (SAE) Press, 1995.

Hartmann, William K. *A Traveler's Guide to Mars.* New York: Workman Publishing, 2003.

Klekx, Greg. *Lost in Space: The Fall of NASA and the Dream of a New Space Age.* New York: Parthenon, 2004.

Lee, Pascal. "From the Earth to Mars." *Planetary Report* 22 (January/February 2002 and May/June 2002): 10–16.

———. "Mars on Earth." *Ad Astra* (May/June 2002): 12–17, 51–53.

Long, Michael E. "Mars on Earth." *National Geographic,* July 1999, 36–51. Photographs by Peter Essick.

Morton, Oliver. *Mapping Mars.* New York: Picador, 2002. An indispensable cultural history of the Red Planet.

O'Sullivan, John. "The Great Nation of Futurity," *United States Democratic Review* 6, no. 23 (1839): 426–430.

———. Unsigned editorial. *New York Morning News,* December 27, 1845.

Ozinski, Gordon. "Hypervelocity Impacts into Sedimentary Targets: Processes and Products." PhD diss., University of New Brunswick, 2003.

Pisanich, Greg, Laura Plice, Corey Ippolito, Larry Young, Benton Lau, and Pascal Lee. "Initial efforts toward mission-representative imaging surveys from aerial explorers." *Proceedings of SPIE* 5297 (May 2004): 106–121. A report on imaging results from two seasons of flights at Haughton Crater given at the Real-Time Imaging VIII conference held by the Society of Optical Engineering.

Pyne, Stephen J. *How the Canyon Became Grand.* NY: Viking, 1998. This is an excellent primer into the cultural geography of the Grand Canyon, and contains an analysis of von Egloffstein's painting of the entrance to Black Canyon and the relationship of romantic art to exploration.

———. "Space: The Third Great Age of Discovery." In *Space: Discovery and Exploration,* 7–14. New York: Hugh Lauter Levin, 1993. Pyne, a notable historian of exploration, has also written a seminal cultural history of the Antarctic (*The Ice,* published in 1986 by the University of Iowa Press) that is relevant.

Raeburn, Paul. *Uncovering the Secrets of the Red Planet: Mars.* Washington, DC: National Geographic Society, 1998.

Sheehan, William. *The Planet Mars: A History of Observation and Discovery.* Tucson: University of Arizona Press, 1996.

Squyres, Steve. *Roving Mars: Spirit, Opportunity, and the Exploration of the Red Planet.* NY: Hyperion, 2005.

Wald, Matthew L. "Mars Mission's Invisible Enemy: Radiation." *New York Times,* December 9, 2003, D1.

Walker, Sean P., and J. Kenneth Salisbury. "Large Haptic Topographic Maps: MarsView and the Proxy Graph Algorithm." ACM SIGGRAPH 2003 Symposium on Interactive 3D Graphics, 83–92. http://portal.acm.org.

Zubrin, Robert. *The Case for Mars.* New York: Simon & Schuster, 1996.

_____. *Entering Space.* New York: Penguin Putnam, 1999. Zubrin expands his version of "manifest destiny" to the rest of the universe.

_____. *First Landing.* New York: Penguin Putman, 2001. A wildly unrealistic, fictional tract based on the ideas contained in the author's previous two books.

———. *Mars on Earth.* New York: Picador, 2002. An oddly biased account of the Mars Society's work on Devon Island.

MARTIAN ART AND LITERATURE/SPACE ART

Belloli, Jay. *Twenty-five Years of Space Photography.* Pasadena, CA: Baxter Art Gallery, California Institute of Technology, 1985. This catalog for an exhibition of the same title includes important interviews between curator Belloli and three JPL image-makers.

Bradbury, Ray. *The Martian Chronicles.* New York: William Morrow, 1997.

Burroughs, Edgar Rice. *A Princess of Mars.* New York: Modern Library, 2003.

Clarke, Arthur C. *The Exploration of Space.* New York: Harper & Brothers, 1951.

Hardy, David A. *Visions of Space: Artists Journey Through the Cosmos.* London: Dragon's World, 1989.

Hartmann, William K., Ron Miller, Vitaly Myagkov, and Andrei Sokolov, eds. *In the Stream of Stars: The Soviet/American Space Art Book.* New York: Workman Publishing, 1990.

Miller, Ron, and Frederick C. Durant III. *The Art of Chesley Bonestell.* London: Collins & Brown, 2001.

Ordway, Frederick I., III. *Visions of Spaceflight: Images from the Ordway Collection.* New York: Four Walls Eight Windows, 2001.

Nasmyth, James, and James Carpenter. *The Moon: Considered as a Planet, a World, and a Satellite.* London: John Murray, 1874.

Richardson, Robert S., ed. *Man and the Moon.* Cleveland: World Publishing, 1961. The illustrations by Bonestell for this collection of nonfiction essays, published prior to the first moon landing, demonstrate his penchant for unrealistic but sublime moonscapes.

Robinson, Kim Stanley. *Blue Mars.* New York: Bantam Books, 1996.

———. *Green Mars.* New York: Bantam Books, 1994.

———. *The Martians.* New York: Bantam Books, 1999.

———. *Red Mars.* New York: Bantam Books, 1993.

The Mars trilogy, with its fourth volume of miscellany as a coda, is not only the most thorough and meticulous treatment of the planet in fiction, but also a fine primer on areology, the socio-economics of post-Fordist exploration, exobiology, and a host of other related topics. Excerpts from the 1997 interview with Robinson in which he discusses his novel *Antarctica* can be found on *Locus* magazine's Web site: www.locusmag.com/1997/Issues/09/KSRobinson.html.

COGNITION AND LANDSCAPE

Appleton, Jay. *The Experience of Landscape*. New York: John Wiley, 1996.

Blaksless, Sandra. "When the Brain Says, 'Don't Get too Close.'" *New York Times,* July 13, 2004, D2.

Fox, William L. *Terra Antarctica: Looking into the Emptiest Continent*. San Antonio, TX: Trinity University Press, 2005.

———. *The Void, the Grid, & the Sign*. Reno, NV: University of Nevada Press, 2000.

Gregory, Richard L. *Eye and Brain*. Princeton, NJ: Princeton University Press, 1997.

Montagu, M. F. Ashley. *Touching: The Human Significance of the Skin*. New York: Harper & Row, 1986.

O'Neill, Maire Eithne. "Corporeal Experience: A Haptic Way of Knowing." *Journal of Architectural Education* 55, no. 1 (2001): 3–11.

Solso, Robert. *Cognition and the Visual Arts*. Cambridge, MA: MIT Press, 1994.

Tuan, Yi-Fu. *Space and Place: The Perspective of Experience*. Minneapolis: University of Minnesota Press, 1977.

INDEX

A

N

R